INDUSTRIAL ENGINEERING AND MANAGEMENT

Industrial Engineering and Management

Edited by

Harinirina Randrianarisoa, Ph.D.

ARCLER PRESS

www.arclerpress.com

Industrial Engineering and Management

Harinirina Randrianarisoa, Ph.D.

Arcler Press
2010 Winston Park Drive,
2nd Floor
Oakville, ON L6H 5R7
Canada
www.arclerpress.com
Tel: 001-289-291-7705
001-905-616-2116
Fax: 001-289-291-7601
Email: orders@arclereducation.com

ISBN: 978-1-77361-231-7 (Hardcover)

About the Editor

Harinirina Randrianarisoa, Ph.D.

Hari Randrianarisoa obtained is Ph.D from Institute of Electronics, Microelectronics and Nanotechnology (IEMN) of the University of Science and Technology – Lille (France). His thesis focused for designing and manufacturing of thermal micro sensors. His interests span from on renewable energy especially solar energy. He has a strong knowledge in technology transfer field. He is currently working as a Business advisor for economic development organization in Québec City and as a consultant Scientist advisor.

Contents

List of Figures ..... *xi*

List of Tables ..... *xiii*

Preface ..... *xv*

Chapter 1 Introduction 1

- Owner Involvement 3
- Complexity 3
- Management 5
- Information Control And Reporting 6
- Time 7
- Safety, Quality And Environmental Issues 8
- Estimating And Risk 9
- Cost Control And Reduction 10
- Terminology 12
- Non-Standard Documentation 13

Chapter 2 Management of Industrial Engineering Projects 15

- Introduction 15
- Project Execution Plan (Quality Plan) 16
- Project Stages 22
- Construction 26
- Initial Operations 27
- Project Risk Analysis and Management 29
- Safety, Environment And Quality Assurance 34
- Insurance 36

Chapter 3 Estimation and Assessment 39

- Introduction 39
- Stages In Estimate Preparation 43
- Estimate Types 44

- Hierarchical Composite Rates 52
- Man-Hour Norms / Standards 53
- Escalation, Exchange Rates And Financing Charges 55
- Contingency Allowance 57
- Risk Analysis And Evaluation 58
- Estimate Content 58
- Interfaces 59
- Control By Estimate 61
- Continued Use Of Estimate 62
- Coding And Control 62
- Estimate Presentation 62

Chapter 4 Planning and Progress Control 65

- Introduction 65
- The Importance Of Planning 65
- Organization 66
- Levels Of Planning 67
- Network Techniques 74
- Critical Path And Float 75
- Progress Measurement 76
- Continuing Control 79

Chapter 5 Cost Material and Document Control 83

- Cost Control 83
- Document Control 99
- Material Control 104

Chapter 6 Procurement of Materials and Equipment 109

- Introduction 109
- Scope of Procurement Activities 109
- Requisitions 110
- The Procurement Cycle 110
- The Procurement Team 111
- Communication 112
- Terms And Conditions 112
- Certification And Certifying Authorities 117

- Material Traceability 118
- Spares And Special Tools 118
- Quality Assurance 120
- Pre-Qualification Of Suppliers 120
- Tender List 121
- Enquiries 121
- Expediting And Inspection 124
- Shipping, Transport And Insurance 126
- Progress Measurement And Reporting 128

Chapter 7 Quality Assurance and Quality Control 131

- Introduction 131
- Dimensions Of Quality 132
- Quality Assurance 139
- Total Quality Management 139
- The Standards 140
- The Systems 141
- Procedures 146
- Team Quality 147
- Sub-Contractual Worker And Supplier Quality 148
- Quality Audit 149
- Quality Improvement 150
- The Way Forward 154

Chapter 8 Coordination of Procedures and Value Management 157

- Introduction 157
- Management Procedures 158
- Design Procedures 159
- Site Procedures 160
- Production of Site Procedures 161
- Typical Site Procedures 162
- People and Procedures 165
- Changes To Procedures 166
- Compliance 166
- Conclusion 166
- Introduction To Value Management 167

- Definition Of Value ... 168
- Vm Definitions ... 168
- Timing Of Studies ... 170
- Value Management Methodology And Job Plan ... 171
- Conclusion ... 175

Chapter 9 **Operational Maintenance ... 177**

- Introduction ... 177
- Planning ... 181
- Estimating And Budgeting ... 187
- Contracting Strategies ... 191
- Control Of Costs ... 194
- Contracts / Financial Audits ... 199

Chapter 10 **Productivity ... 201**

- Introduction ... 201
- Purpose Of Productivity Improvement ... 202
- Different Approaches Lead To Different Results ... 205
- Drivers Of Profitability Development ... 208
- Engineering Approach For Productivity ... 209
- Three Levels Of Improvement ... 212
- Three Dimensions Of Productivity ... 220

References ... 237

Index ... 243

List of Figures

Figure 1.1 Project flow chart (the joint development board sponsored by the association of cost engineers and the royal institute of chartered surveyors, n.d.)

Figure 1.2. Potential for savings and extra costs in project life cycle (the joint development board sponsored by the association of cost engineers and the royal institute of chartered surveyors, n.d.)

Figure 4.1. Level 1 bar chart with key dates (milestones) and progress curve (the joint development board sponsored by the association of cost engineers and the royal institute of chartered surveyors, n.d.)

Figure 4.2 Level 2 bar chart programme (the joint development board sponsored by the association of cost engineers and the royal institute of chartered surveyors, n.d.)

Figure 4.3. A job card (the joint development board sponsored by the association of cost engineers and the royal institute of chartered surveyors, n.d.)

Figure 4.4. Usual curve of level 2 progress (the joint development board sponsored by the association of cost engineers and the royal institute of chartered surveyors, n.d.)

Figure 5.1. A typical work break down structure (the joint development board sponsored by the association of cost engineers and the royal institute of chartered surveyors, n.d.)

Figure 5.2. A typical code of account (the joint development board sponsored by the association of cost engineers and the royal institute of chartered surveyors, n.d.)

Figure 5.3. Productivity factors for predictions (the joint development board sponsored by the association of cost engineers and the royal institute of chartered surveyors, n.d.)

Figure 5.4. Curve of consumption and commitment (the joint development board sponsored by the association of cost engineers and the royal institute of chartered surveyors, n.d.)

Figure 8.1 Review levels explained (the joint development board sponsored by the association of cost engineers and the royal institute of chartered surveyors, n.d.)

Figure 8.2. Oppurtunity and change cost dependence in projects (the joint development board sponsored by the association of cost engineers and the royal institute of chartered surveyors, n.d.)

Figure 8.3. A fast diagram (the joint development board sponsored by the association of cost engineers and the royal institute of chartered surveyors, n.d.)

Figure 10.1 Improvement spread (sakamoto, n.d.)

Figure 10.2. Design approaches (sakamoto, n.d.)

Figure 10.3. Uprising bf ratio (sakamoto, n.d.)

Figure 10.4. Dts result on an assembly (sakamoto, n.d.)

Figure 10.5 Pile-up and decomposition methods (sakamoto, n.d.)

Figure 10.6. Accumulated chart (the joint development board sponsored by the association of cost engineers and the royal institute of chartered surveyors, n.d.)

Figure 10.7. An accumulated chart use (the joint development board sponsored by the association of cost engineers and the royal institute of chartered surveyors, n.d.)

Figure 10.8 M and p meaning with time standard (sakamoto, n.d.)

Figure 10.9. Methods against standard time (sakamoto, n.d.)

Figure 10.10 Cost from raw materials for finishing (sakamoto, n.d.)

List of Tables

Table 3.1. Estimate exactness (the joint development board sponsored by the association of cost engineers and the royal institute of chartered surveyors, n.d.)

Table 10.1. M, p, u and travel time (sakamoto, n.d.)

Preface

Since the early start of mechanical uprising, manufacturers are reliably attempting to enhance the qualities of their manufactured products and proficiency of the manufacturing processes. Man began manufacturing from very basic level to deliver extremely simple products to make his life less difficult. There were no machines and gadgets, all the manufacturing work was finished by hand and the standard things of daily utilization, for example, hatchet and sledge were utilized as devices for help with manufacturing operation. As, time passed, man found numerous new wonders of nature, the information of science, technology, designing, engineering and administration advanced and deliberate studies on manufacturing were begun. Today, Manufacturing has developed to an enormous level, it has turned into a tremendous subject.it has turned into a many disciplinary approach which is a combination of numerous sciences, designing, management, business and so forth.

The latest couple of years have witnessed matchless changes all through the global industrial economy. Speedy changes in the business divisions demand assortment of item and product life cycles together with first class items at low expenses. Customers now request astoundingly elevated amounts of value and enactment. This worldwide trend and market rivalry has spurred industrial firms to scan for dynamic and motorized manufacturing systems in all fields. Thus, many manufacturing procedures, innovations and frameworks have come in presence and research is as yet preceded in this field and some new, drastically extraordinary optimized manufacturing procedures, advancements and frameworks are predicted to appear in the coming years.

The purpose of this book is to give a depiction of present day industrial engineering and industrial management practices with novel procedures which are extensively playing a pivotal role in the industrial procedures and frameworks. This book introduces a snapshot of the all-around prospecting industrial world alongside the most recent market issues and how they are being tended to.

This book is of interest for graduate level audience who are keen to get in touch with industrial engineering systems, processes and technologies. This book is also of interest for manufacturers and technical staff working in today's innovative industrial corporate world. The first chapter offers a general sight of industrial engineering and management, processes, technologies and techniques which are used for monitoring them. In the later chapters Management of industrial engineering projects, Estimation and assessment of engineering projects, Planning and progress control, Cost, material and document control, Procurement of materials and equipment, Quality assurance and quality control, Value management, Operational maintenance and Productivity have been explained.

Editor

CHAPTER 1

Introduction

Industrial Engineering and construction is a broad articulation covering the outline and construction of capital projects for a wide spread of construction and energy related industries including:

- Gas and oil;
- Chemical and process;
- Refining;
- Nuclear;
- Power production;
- Iron and Steel;
- Pharmaceuticals;
- Oxygen and different gasses;
- Food and drink;
- Water and sewage treatment

Some of these ventures are indigenous and have built up their practices with the construction sectors of other nearby industries, by adapting the same basic terminologies of administration and control while others started in different parts of the world.

Current practice has in this way been affected by the nature of the projects normally attempted and the conventional practices in the area of the world where the industry has its foundations. In addition, owners and contractual workers in the above areas are frequently among the world's biggest companies, numerous of which have made big investment both regarding experience and construction costs in their specific technique for working together. (The Joint Development Board sponsored by the Association of Cost Engineers and the Royal Institute of Chartered Surveyors, n.d.)

Although there are historical beginnings numerous of the engineering sectors handle capital projects in a same way because of various normal factors. These include:

- Level of owner involvement;
- Remoteness of area;
- Complexity of the projects;
- Coordination of high worth projects, with huge multidiscipline,
- Contractual worker teams;
- Time limitations;
- Lack of standard documentation;
- Dangerous nature of the plants being developed;
- Level of hazard included;
- Preparing of the construction profession

The different elements that impact the economics of the finished plant will have been considered in the possibility study about before the project gets to proceed. The plant might be competing with comforts assembled a few years previously which have their costs recorded, and the product from the new plant must match these current costs by the utilization of new innovation to increment productivity, by reducing working and upkeep costs, and by guaranteeing the most reduced capital cost for the plant. This book actually deals with the management methods and procedures important to get this economic success.

The project flow chart, Figure below, shows the stages which are typical of an engineering capital project from idealism to completion.

OWNER INVOLVEMENT

Numerous factors affect the way in which a related owner of a plant deals with his design/construction team. The distinctive owner involved in an engineering project capable to have much technical experience and expertise than his opposite number in other development industries in that he:

- Might have same plants anyway;
- Will frequently be the end user;
- May have design standards and in-house design plan ability;
- May hold a procedure license;
- May have a project administration capacity including requirement,
- Cost organization and contract specialist will every now and again keep up a supervision and coordination part;
- Will surely know how the plant can best be worked

The typical owner of a engineering project will be more involved in the everyday choices and take a far more greater level of duty regarding the project than will most owners in the engineering and structural designing projects. This contribution has a greater impact all through the entire association of the work and on the legally engineering connections between the different gatherings.

Generally, schemes take long time to execute and at first a owner must consider plant limit, development costs, generation openings, feedstock, energy, transport or more all the available market and the value the product will order later on. This will obviously call for examine with respect to what different associations are engineering offices, or wanting to develop offices, which would influence the product market or cost. (Adedeji Badiru, n.d.)

COMPLEXITY

Many designing projects are by their exceptionally complex nature and incorporate exceptional plans or procedures. Owners request the most current advancements either to create another product or to guarantee financially satisfaction and so on and as often as possible require the completed product in a short improvement period. This requires parallel working between the design and development stages and the requirement for multi-contractual construction side with an ensuing potential for day, disturbance and related additional costs.

The design of a plant all in all may fuse designs given by different authority suppliers, who can't design their component in isolation from the main design, yet whose design is required by that main design. The subsequent iterative methodology requires close control and assisting of data that streams between the different gatherings, all together that all the different designers, specialist suppliers and construction contractual workers work from the most a latest data constantly. (George A. Steiner, n.d.)

The time required to make a few things of modern equipment may require that they are requested in front of any agreement being offered for their establishment. Moreover, mass materials, e.g. pipe, valves, electrical and instrumentation links, and so forth, Might be acquired by the owner or the owner's managing contractual worker for reasons of similarity or purchasing power. This offers rise to construction contracts where a few or most of the materials are given as 'free issue' to the contractual worker, which expects controls to be introduced into the way in which those materials are acquired, issued and utilized.

On expansive projects, it is doubtful that any one fundamental contractual worker will have all the essential abilities and assets expected to create and construct whole project. The outcome is the need to discuss site access to, accessibility of design and the free issue of materials and equipment, with the parallel or consecutive working of various significant construction contractual workers, sub-contractual workers and equipment makers.

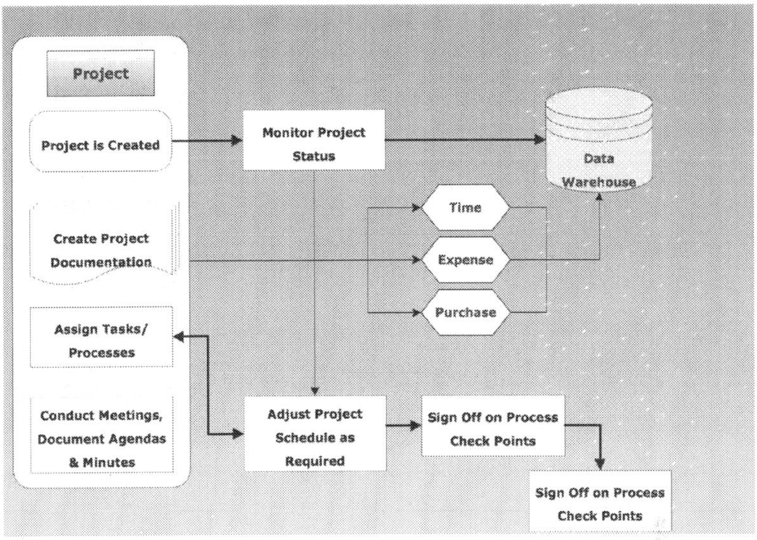

Figure 1 1: Project flow chart

Following construction, the plant will be pre-authorized lastly authorized and trial to confirm the plant's execution. Some of this work might be attempted by the owner, some by the administration contractual worker and some by the construction contractual workers, sub-contractual workers or equipment providers.

The effect of design versatile nature, the coordination of the design, acquisition and construction exercises and parallel working, is a notable administration work which features the requirement for tight control in wording of cost, time and data.

MANAGEMENT

At the point when approval is given for a large project to continue, many individuals are utilized for quite a long time or years and a large number of pounds (or dollars, dinars, and so forth.) are consumed. The accomplishment of such a project once in a while relies upon a single person. In spite of, one single insufficient individual at an adequate level of impact can bring disappointment, overspend or delay (regularly every one of the three).

The very term 'extend administrator / manager' prompts a misconception of the part of supervisors, and in reality it supports the conviction that undertakings can be "oversaw" by an individual, though they must be controlled by a group. The extend supervisor is the leader of the group and accordingly should have satisfactory information of the engineering issues, good directions and the law as it identifies with the project, together with plenty of sound judgment, the aptitudes of a representative, a capacity to tackle disgusting issues rapidly and enough regard from the group to get greatest action and collaboration from them by any time.

INFORMATION CONTROL AND REPORTING

Understanding the significance of appropriate announcing is basic if a project group is not to be covered by the plenty of data produced by an average significant project. Expansive projects are excessively confusing, making it impossible to permit extend administrator / managers to consistently do any of the point by point work themselves, yet a project administrator requires important data to be given in a consolidated shape keeping in mind the final goal to keep up a review of the project and its key exercises. (Tonchia, 2008)

The project will subsequently need a modified administration to give intelligible data in report frame, bolstered by additionally definite levels of data accessible to be analyzed, adjusted and followed up on as required.

Unless the conveyance of the expansive measure of data and reports produced by a project is controlled by a system which ensures that each part gets just what is required, the project administration group can be overpowered by pure volume of printed material containing data which is continually changing and creating. In addition, unless such data is coded and effectively documented, the likelihood of its fruitful recovery is remote.

Data control is thereby a notable action on building projects, not exclusively to guarantee that all the different projects individuals are working from the most upcoming documentation and information yet additionally to guarantee that drawings and different records are accessible to the owner which precisely mirror the as-constructed condition, since once finished the plant is a continuous operation requiring upkeep and believable future advancement.

TIME

Time on a project is frequently extremely essential, given the need to meet platform desire and client request; and now and again the prior acquiring limit will exceed any expanded cost of quickening the design and construction process. Time on all projects turns out to be progressively basic as a greater amount of it passes. At the point when the project is in the possibility arrange, a extend completion and charging date maybe five years later on may appear far ahead; however when endorsement to continue is at last got a year can have slipped by, however the finishing date seldom moves. On the off chance that the project turns into the subject of natural complaints additionally delays can collect. (Tonchia, 2008)

Despite natural issues, the advancement of the design is the primary phase of the project to feel the expanding weight applied on it by the progression of time; however the proposed final date may at present show up a long way off.

Additionally pressure is applied on the design group as the acquisition movement requests design criteria and execution necessities all together that tenders can be looked for the buy of long lead things. There ik now an expanding mindfulness that time is undoubtedly breathing easy left to complete and commissioning is somewhat tight.

At the point when finally manufacture and construction begin, time may have the entire project group by the throat. The time required for the sensible successive movement through the work may never again be accessible. Design may have overcome and acquirement be postponed as an outcome; however the end date remains cast in stone. The results of this are not hard to anticipate. (Waldemar Karwowski, n.d.)

At the point when the program slips, the reasons are frequently different or not particular. The need to quicken the program more often than not requires promote consumption by the owner well beyond the first contract which, as anyone might expect, he may not be excessively sharp, making it impossible to cause. There will be a compulsion to postpone unpalatable choices with the expectation that consequent occasions will make up any past deferrals to the program.

The general impact of a deferral to an owner might be loss of income and conceivable loss of business sectors and the owner should adjust his alternatives in choosing whether to acknowledge the postponement or bear the cost of quickening the work.

The significance of the arranging specialist's capacity in guaranteeing that the work is finished on time can't be overemphasized. The numerous exercises expected to design the work, secure equipment and materials, grant contracts, arrange contractual workers and providers, develop and commission the plant and set it to work should be painstakingly sequenced and advance observed, and healing moves should be made quickly to overcome issues.

The arranging designer will be confronted with the troublesome assignment of producing an general project program demonstrating what is required, of whom and when, and after that guaranteeing that all different gatherings required in the venture work to plans which are perfect with this ace program.

SAFETY, QUALITY AND ENVIRONMENTAL ISSUES

Safe construction and safe operation is the subject of various Demonstrations of Parliament and quality confirmation is presently by and large utilized all through the entire of the designing business. So also, the becoming ecological campaign has additionally made owners and contractual workers alike more aware of the need to work in a naturally worthy way.

While comfort and security is especially essential to the engineering businesses in perspective of the risk fundamental in any major modern plant, if the in demand quality is not accomplished support costs, assurance of generation and natural insurance would all be able to be unfavorably influenced. (Bessant, n.d.)

It is presently frequently the case that a working or assembling organization will request that any contractual worker or sub-contractual worker utilized on a venture must have quality affirmation and quality control (QA/QC) systems predictable with the necessities of IS0 9000, and that they likewise have an ecological approach. In the event that an organization does not have such controls inside, at that point it is probably not going to be considered when delicate records are readied.

Natural contemplations are urgent to the program. On the off chance that the project is the subject of natural protests amid its improvement and goes to open request, this may take year and a half, and a legal audit will take a comparative period, while arranging assents or emanation licenses will take around a month and a half each time the project goes to council. On the off chance that, toward the end of this period, another site must be chosen and obtained, this procedure will begin once more, with disastrous consequences for the program.

ESTIMATING AND RISK

Many designing undertakings are of high capital worth, and are perhaps done in remote areas; many utilize forms at the points of confinement of current innovation, and many are one of a kind or adequately not at all like each other to make it hard to utilize a standard, abnormal state, cost database.

Thus assesses are required to be created to a far more notable degree of detail than in other construction project with the final goal for them to be adequately adaptable to meet the requests of every one of a kind domain and to empower the project director to recognize, break down and nearly deal with the critical dangers required through the project time frame.

Expansive undertakings have unique components which make them extraordinary and, at the point when things turn out badly, more costly. Reasons can fluctuate however incorporate the following:

The administration of significant projects is regularly embraced by modified groups of experts united for that one anticipate and dispersed, after success, with resulting loss of group connections what's more, involvement of cooperating. (Horngren, n.d.)

Extraordinary things of plant might be requested months, some of the time years before establishment. Any oversights or postponements in the arrangement of this plant can't be settled by an pleasure trip to the nearby developers' vendor to buy extra amounts or a substitution. The necessity to utilize huge quantities of authority fitters, welders, and so on can drain the assets of the region, requiring the examination of the different choices for off-site manufacture, importation of work, and so on. Isolate controls and pro providers might be designing their commitments in separation, from now on the likelihood of mistake or change increments unless great coordination is drilled.

Designing projects are site-particular, modified components and in some cases fuse untried innovation. This rundown may have the impact of putting the greater part of the contractual workers and group individuals, however experienced, on an 'expectation to absorb information' at all phases of the work.

In spite of the fact that there will dependably be a drive to enhance administration and project controls, the critical dangers related with designing activities can't be completely killed. It is generally perceived in the engineering industry that hazard assessing procedures are basic and effective apparatuses, especially amid the early periods of a project. They are in this manner used to help the comprehension of the dangers and to recognize regions where administration consideration is most expected to moderate their belongings. The dangers for the most part emerge from the unpredictability of the plant, the remote nature of the run of the mill site, the earnestness with which the plant is required to be finished and the zones of new and untried innovation, all of which will test the most experienced estimator.

In spite of the fact that the point must to dependably be to enhance execution and utilize 'fit for the reason' designs and particulars, the choice to diminish cost by lessening determination must be taken against the foundation of a sensible gauge which incorporates a legal thought and assessment of the project dangers. To acknowledge late in a project that elements that were a piece of the unique brief however not allowed to decrease cost could have been thankful inside the first spending won't charm the estimator to associates or manager.

COST CONTROL AND REDUCTION

The truth of a project will imply that unless powerful cost control is practiced expenses will heighten and the activities of the most experienced of estimators utilizing the absolute best of data will be wasted.

Control is not the separate part of just observing and representing cost, nor is it restricted to the cost of material and construction. Cost control is a dynamic part, which starts on the very first moment of the project, with the control of administration and design, and it proceeds through acquisition of materials and construction of the plant, to consummation and settlement of all accounts. Cost control is embraced by fetched engineers whose capacity is to guarantee that the work is embraced in the most financially know-how by looking for monetary arrangements, checking use, dissecting execution, recognizing issue regions and prescribing preventive or curing activity. The position of cost specialist might be filled by one who is by calling a cost specialist or amount surveyor; however the term amount surveyor is not broadly utilized as a part of the building enterprises. (Horngren, n.d.)

Organization expenses can be decreased by different means including disposal of waste, evacuation of extra necessities or by expanded effectiveness. Expenses can likewise be diminished by decreasing ability or particular, however regardless of whether this constitutes a "sparing" might involve sentiment since the owner is getting less for his cash.

Waste to be dispensed with incorporates poor execution, absence of coordination what's more, change for its own purpose, all of which causes upset working and delay. The utilization of non-standard things, be they things of equipment, contract documentation, methods or affirmation necessities, as opposed to an 'off the rack' fit for the reason, again acquires surplus additional cost or hazard.

The CRINE (Cost Decrease Activity for the New Time) report features a hefty portion of these components regarding the seaward oil and gas businesses and stresses the investment funds that it considers could be pleased in if the conventional severe types of agreement were supplanted by a helpful connection between contracting parties.

The truth will surface eventually whether the goals of the report and the Latham Report into the UK construction industry can be converted into the savings demanded and whether comparable activities are similarly legitimate in other construction situations.

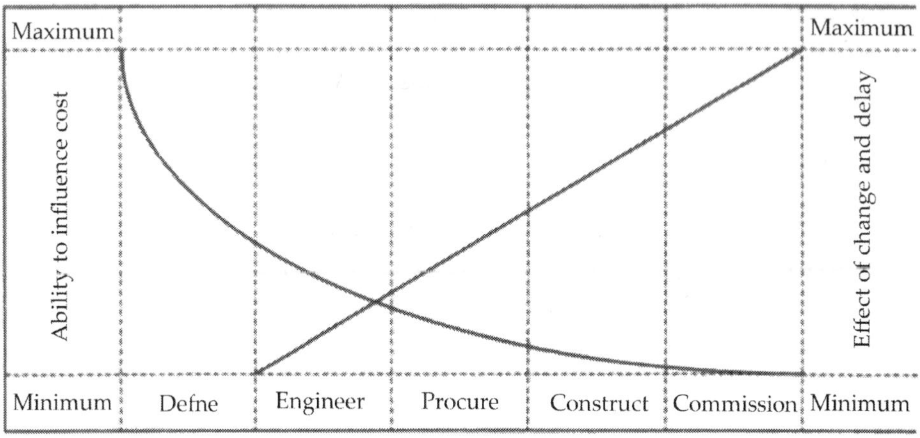

Figure 1 2: potential for savings and extra costs in project life cycle (The Joint Development Board sponsored by the Association of Cost Engineers and the Royal Institute of Chartered Surveyors, n.d.)

It is now and again important to challenge whether something saw as fundamental to a project is in reality energy. The expulsion of pointless substance, or over specification, from a project is a real sparing and the utilization of the strategies of significant worth designing portrayed in Section 4 challenges each part of a project and the segments inside that project.

Value engineering is a method which requires a deliberate survey of every component of the project and of each thing of equipment inside the project, addressing reason and cost with a specific end goal to distinguish reserve funds. (Salvendy, n.d.)

Figure 1.2 demonstrates that while the capacity to make reserve funds lessens as the extend advances, the impact of progress and postpone increments. This obviously shows that to get best impact any cost sparing audits must be begun tolerably early. It likewise shows that, amid this same early period, administration structures, controls and techniques must be set up to enhance and keep up execution and maintain a strategic distance from the total impacts of progress, delay, modify and disturbance.

TERMINOLOGY

The absence of institutionalization inside the engineering is obvious when taking a gander at a rundown of clearly compatible terms which, best case scenario cause misconception and even under the least favorable conditions legally binding blunder, question or other trouble. The need to completely comprehend the wording utilized is self-evident. A term evidently in like manner utilize might be given diverse definitions on various project. For instance, in this book the individual for whom the project is being attempted is alluded to as 'the owner' while somewhere else the terms 'engineering', 'customer', or "administrator" are regularly being used. So also the non-specific term "change" is utilized to represent the accompanying terms which are all in all utilization to represent a change to the design, detail, methods for construction, or timing of the work:

- Specialist's direction;
- Change order;
- Pattern notice

Much more prominent challenges identify with the term, 'project manager' which is utilized in this book to represent the individual in general charge of the project. A few owners utilize the term project manager, while others allude to 'the designer, as do some distributed standard states of agreement.

To add to the confusion, a few states of agreement require both the owner and contractual worker to pass on extend directors, while others may name a project supervisor to be responsible for the general project, however utilize types of contract which allude to the individual at last in charge of an agreement as being 'the engineer' or the 'owner agent'. Contractual workers and sub- contractual workers may thusly name a project manager for their parts of the entirety. (Sakamoto, n.d.)

In considering choices by and large utilize it is in this way energy to consider the project all in all as well as the different contracts set inside that project. The capacity of a specific individual or record should not be accepted without reference to a more exact definition impossible to miss to that project.

NON-STANDARD DOCUMENTATION

An important component significantly affecting organization inside the industry is the absence of institutionalization of agreement archives, regulatory practices, strategies for measurement etc. This absence of institutionalization influences the productivity of the engineering in different ways:

It lessens the likelihood of gathering and trading information between venture projects and organizations. It causes blunder in the comprehension of fundamentals and requires tenderer to inquiry through documentation to discover potential legally binding perils and contractual workers to fall foul of conditions which they had misconstrued. It expects work force to familiarize themselves with the frameworks and methods for each project.

CHAPTER 2

Management of Industrial Engineering Projects

INTRODUCTION

In all attempts collaboration and participation are key fixings to the achievement of designing undertakings. An owner will depend on his venture administration group; they thus will be dependent on contractual workers, sub-contractual workers and providers who have been decided for their experience and reasonableness. In any case, every individual from this more significant group is reliant on others and the techniques, frameworks, contracts and controls must permit the advantages of participation to come through. Such an adjust is best accomplished through a valuation for the prerequisites of different individuals from the venture and it is trusted that this book will enable each project part to see how alternate controls do their work and in this manner help them in accomplishing the collaboration and effectiveness we as a whole attempt to get. Owners and contractual workers required in mechanical designing activities are continually looking for better approaches for undertaking capital activities with a specific end goal to increment the effectiveness of the engineering, to enhance the financial aspects of working and to adjust to the changing security and condition prerequisites around the globe. (The Joint Development Board sponsored by the Association of Cost Engineers and the Royal Institute of Chartered Surveyors, n.d.)

In this manner, while each exertion has been made in this book to guarantee the accuracy of the data exhibited, the Joint Improvement Board can't assume liability for any sincere beliefs, mistakes or superseded data since statutory prerequisites are always showing signs of change and practices and techniques differ from organization to organization. Any individual taking a place of duty inside a building extend is firmly encouraged tō familiarize himself with the general necessities of a specific organization and the particular prerequisites of a specific extend since no book can cover the large number of contrasts in style and definition being used around the world.

PROJECT EXECUTION PLAN (QUALITY PLAN)

Because of the complicated and large size of many engineering project ventures, the quantity of people included and the assortment of their commitments, the need for collaboration must be underlined and a domain made in which a proficient group can develop.

The project execution plan is the zenith of the project quality framework; it expresses the wide range prerequisites of the project and makes an interpretation of these into a point by point and sound procedure. Moreover, it starts different archives and documents that grow, execute or demonstrate the aftereffects of what is depicted in the detailed document and report. (Adedeji Badiru, n.d.)

Project execution plans are delivered by the project group to cover its fundamental exercises. Comparable execution plans will be delivered by each real contractual worker for their exercises that will match with the project execution plan arranged by the project group.

Development of the project execution plan

The first write-up of project execution plan is typically created amid the conceptual planning and design stage with the last project execution plan being issued before beginning of detailed design. This expresses the fundamental targets of the project group and the techniques, project administration processes and exercises important to the accomplishment of these primary project destinations. For the most part it won't state how the processes or exercises ought to be conducted, as these will be created in an independent and more detailed document and report. (Adedeji Badiru, n.d.)

The project execution plan should reference all Nitti gritty scheduling and planning documents, for example, the acquisition and contract design, a firm's organizational chart, and so on. These portray how the project is to be created, the assets required, and organized with the goal that the project team can gauge its execution against the project execution plan.

The project execution plan is the principle technique and strategy document for the project furthermore, it should plot the fundamental structure of the project including:

- Project targets;
- Project administration
- Project association and organization
- Project program
- Project control methodology
- Quality designs
- Security and ecological arrangement
- Acquisition methodology
- Contracting methodology
- Spending plans
- Asset methodology

Where an organization has started to work on a mission and vision explanations, the project execution plan will likewise contain such proclamations that are pertinent to the project.

Association and organization charts inside the project execution plan must recognize the positions to be hired, their roles, parts and revealing connections. Individual and group obligations and specialist levels ought to be plainly characterized as ought to be the divisions of work and interfaces. Nitty gritty sets of expectations and personnel extents and scopes of work may likewise be required.

Project objectives

One of the main considerations in setting up a decent group of individuals for a project is to center consideration on the coveted come about by giving a comprehension of what will be thought to be a fruitful project. The project objectives are an announcement of what must be accomplished if the project is to be viewed as a win. The project goals ought to give a concentration towards the essentially critical parts of the project and ought to either be a measure of its prosperity or ought to support the conduct prompting that win. Regular project objectives comprise of operational prerequisites, project fulfillment milestones, health and safety objectives, quality of the product being produced and costs involved. Project objectives ought to be sensibly little in number, such that every one of the objectives can be recollected and be viewed as essential. They ought to comprise of clear, unambiguous, vital straightforward explanations that can be easily understood and distinguished by the project group and the individuals who bolster it. Goals ought to be practical and achievable; the inability to accomplish impossible objectives quickly devastates assurance and morale of project group and rapidly disappoints them. (George A. Steiner, n.d.)

Mission and vision statements

A few firm business owners and contractual firms set mission and vision statements for their organization or project over the level of project targets. Such statements include a short, sharp, exact statement of goal that goes about as a core interest towards which the objectives and basic achievement factors are coordinated.

The statement of purpose / mission statement is 'the primary concern' and is an announcement of what the organization, or project is trying to accomplish in "hard" terms, i.e. 'to accomplish something by a set date or in an expressed cost'. The vision statement is a 'clear picture' of a driven, attractive future that is preferable in some essential path over the present state, i.e. 'to be seen by clients as being best in class'.

Critical success factors

Fundamental achievement factors are those which portray how the project targets and objectives will be accomplished. Some basic achievement factors which are basic to all projects incorporate, for instance:

- Getting input contribution from the operations team before initiating subtle element design

- Arrangement of a safe and protected workplace
- Flourishing a high performing coordinated project administration group
- Fulfillment of design of an area before beginning construction

Each project will have remarkable conditions subordinate upon the essential targets of the project which may make some generally acknowledged factors unimplemented.

Project programme and milestones

Planning is the sequencing, arranging, checking and reporting of various project events, exercises, points of reference and assets and gives the datum against which the project status, advance and alterations can be gauged.

The project programme consolidated into the project execution plan will in a perfect situation be the level 2 program. Points of reference or deadlines recognized in the project plan check basic project events, for example, key choices or endorsements, or vital interfaces between subordinate exercises and core interests and special consideration on the project focuses. (The Joint Development Board sponsored by the Association of Cost Engineers and the Royal Institute of Chartered Surveyors, n.d.)

Procurement strategies

Since high extent of the project sum is used on materials and job work, the establishment of a sound acquisition and contracting procedure is a standout amongst the most basic components in the project execution plan.

Procurement and acquisition is normally controlled against an procurement plan for that recognizes all materials and equipment to be acquired by the team delivering the plan and conceivable resources of materials. Specific reference will be made to things which go under the prerequisites of EU and UK directions administering rivalry and offering. The procurement plan is additionally a report against which acquirement exercises are observed and controlled.

Free issue equipment and materials

Situations on a project may direct that specific materials (a terminology proposed here to incorporate equipment) are to be issued to the contractual worker by the proprietor on a free issue premise, eminently vessels, electrical and mechanical equipment, instruments, and so on together with mass things, for example, pipework, electrical link and so forth.

The choice to give materials on a 'free issue' premise is made for different reasons including:

- Design data being required from the expert provider in request to finish the general design
- Things being of a pro sort and requiring exceptionally exact models of execution and detail
- Things being on a broadened conveyance and the contractual worker being not able to acquire the materials in a period foreseen by his construction program for that component of the work
- The purchasing power of the proprietor
- The necessity to coordinate materials or equipment as of now being used on an
- Proprietor's current establishments; the proprietor possesses or manufactures the material.

Where a provider is required to design the equipment, assention must be made with regards to the documentation to be delivered, the program for its production and the endorsement cycles required. Such data is indispensable to the essential design exertion and interfaces between the provider and designer must be overseen successfully if the design is to continue in an effective way.

Supplier (vendor) engineers

With regards to this discussion a contract strategy is the separating of the project into bundles of work and settling on the most appropriate technique for funding of each bundle.

The idea of the work bundles is reliant on various elements comprising:

- Contents of the general construction program. Is it important to initiate site cleaning in front of end of the primary design?

- Closeness of the work with different parts of the project. Are there a
- Number of comparative 'structures', how would they relate geologically and
- Regarding program, would they be able to be consolidated into a solitary bundle?
- Single point obligation. Is it conceivable to keep away from illusions and disputes by making a single contractual worker at full in charge of the work?
- Status and accessibility of design at enquiry and honor stages and amid the development time frame. Is a singular amount contract conceivable, is accessibility of design prone to cause continuous issues of postponement and disturbance?
- Early specialist contribution to design. Is it important to hire a contractual worker in stage of design being made with a specific end goal to give a contribution to the design on manufacturability and constructability, and so forth?
- Advantages of scale. Will the "overhead" expenses of site administration and plant usage, and so forth for a vast bundle deliver more prudent profits than various littler bundles?

Once the idea of the different contract work bundles has been chosen the sort of contract and agreement will be assessed.

The sort of contract agreement and the way in which the contractual worker is to be paid relies upon the idea of the work and on the status and accessibility of design and materials at enquiry, at grant and amid the development period-can a single amount be asked for or will the probability of impedance (change, defer disturbance, and so on.) from other parallel gets, the proprietor or his designer, require the utilization of a reimbursable type of agreement?

The control of the readiness, bidding and grant of contract agreements is against a contracting plan which is like the acquisition plan in that it distinguishes the work to be contracted (or subcontracted) by the group, together with data identifying with program requirements and to the foreseen worth also, kind of agreement to be utilized, e.g. single amount, measure and worth, and so forth. Similarly as with the acquirement plan specific reference is made to things which go under the necessities of EU and UK controls. (Tonchia, 2008) Forthcoming contractual workers might be recorded in the contracting plan and any screening prerequisites recognized, with a remittance made in the program for this prequalification process.

PROJECT STAGES

From early times, projects have been separated into a progression of stages. Proprietors utilize different terms for the phases of a project and each project builds up its own particular meanings of the required phases, which may regularly include:

Design:

- Practicality or feasibility phase
- Conceptual design
- Preparatory design
- Detail design and acquisition
- Design results and deliverables

Construction and fabrication

Introductory operations:

- Pre-commissioning;
- Commissioning;
- First start-up;

Operation.

A project may consolidate or subdivide the above stages to suit its specific prerequisites.

Feasibility stage

The feasibility phase of a project will contain studies attempted on a exceptionally summed up premise to assess the specialized technical and business benefits that the plant will give to the proprietor.

Conceptual design

Conceptual design might be done within the firm's office by staff designers maybe supplemented by some staff or by pro design temporary contractual workers.

In spite of the fact that the best effect on conclusive cost can be made amid the conceptual design phase, proprietors and their counselors over and over again design this stage inadequately and designate deficient time and monetary and HR to these crucial early phases of a project. Stress to start a project rapidly can bring about consequent inefficacy and expensive blunders. (George A. Steiner, n.d.)

Amid the conceptual design phase any new innovation is recognized, ponders made, remarkable data incorporated and the design theory created. The choice to consolidate and utilize new innovation and technologies can extraordinarily influence a project and should be a harmony between potential cost reductions inferable from new innovation and the danger of likely cost invades because of their incorporation and development.

The reason for the Conceptual design phase is to:

- Affirm that all the engineering arrangements proposed are in fact also, monetarily practical;
- Guarantee all basic specialized technical issues have been recognized and tended to;
- Look for restricted arranging assents;
- Plan and schedule preparatory engineering

Preparatory design

The points of interest of the conceptual design are evaluated by the group and their inclinations characterized and accepted, or generally disposed of from thought.

The operations team's significant share to the project will happen amid preparatory design phase. Amid this stage the operations team ought to survey and support all items to be delivered. As the project advances through preparatory engineering the consequences of the feasibility phase and conceptual design phase are integrated into equipment and bundle specs and the premise of design created for detail engineering. (The Joint Development Board sponsored by the Association of Cost Engineers and the Royal Institute of Chartered Surveyors, n.d.)

Ability of the facilities to be constructed is confirmed amid preparatory design. Contractual workers and particularly seaward module fabricators have distinctive strategies or successions of construction. Where constructability is probably going to be a problem, potential contractual workers ought to be welcome to audit the design also, their remarks considered for integration in that. (Adedeji Badiru, n.d.)

In certain extraordinary cases, i.e. where a specialized technology is required or, on the other hand where early culmination is foremost, it might be important to choose a contractual worker before the fulfillment of preparatory design. A contract agreement granted ahead of schedule in the design stages won't have an all-around characterized extent / scope of work and is in opposition to the customary and suggested routine with regards to granting contracts in view of a completely created work scope. The arrangement to this issue may require some imaginative deduction in the contracting procedure.

The motivation behind preparatory design is to:

- Activate and mobilize the project group, design group and project administrations
- Direct examinations to improve and characterize design necessities;
- Increment the exactness of scheduling, planning and cost data
- Lead health, safety and environmental investigations
- Create process and instrumentation graphs, single line outlines and formats;
- Examine constructability
- Affirm new or creative innovation
- Orchestrate arrangement of provider information, and put in purchase requests as important to meet the program
- Look for endorsements and planning assents.

Administration of augmentations and changes turns out to be more imperative as the design advances and the project group will utilize a formal method of change control amid preparatory design. The expanding significance put on the control of changes to the design, their cost and impact on the program, serves to underline why alternatives ought to be considered in full amid the prior stages.

Detail design

Change control turns out to be much more crucial amid detail design.

Changes ought to be stayed away from wherever conceivable, yet in the event that progressions must be made, they ought to be tested and if considered adequate ought to be integrated when they are found with the goal that their impact can be limited. Changes may incorporate changes to the design, as well as changes to program, cost, conveyance times, and so on.

The exercises amid detail design are to:

- Create subtle elements and conclude drawings and specs for construction and start-up work
- Get materials, plant and equipment and construction contracts
- Guarantee design follows any safety case prerequisites and environmental evaluation
- Acquire confirmation assntion.

Detail design will every now and again be embraced by an engineering contractual worker, with the proprietor's part being to stipulate project prerequisites, to support indicated key parts of the design, to distinguish the operational necessities and to deal with the interfaces with outside gatherings, for example, co-venturers' and government organizations. The proprietor's part ought not be to stamp the contractual team man for man, or to check the contractual worker's work in detail, or to lead the design phases. The proprietor's part rather ought to be to guarantee that his necessities identifying with quality, execution, operability and safe practice are met, while enabling the engineering contractual worker to utilize his experience and aptitude in the productive design and development of a perplexing project. This necessity makes it fundamental that accomplished and skillful contractual workers are chosen. (Adedeji Badiru, n.d.)

Design deliverables

Each phase of a project and each work bundle inside a project will be required to create something unmistakable, i.e. designs, general plans, specs, enquiry archives, equipment and materials on site, process guarantee testaments, and so forth. The physical consequences of design and other administration exercises are often alluded to as deliverable. It is the deliverable that give the premise to revealing project design status and for perceiving accomplishment.

CONSTRUCTION

Construction will be completed by at least one fundamental contractual workers each of which will utilize different authority sub-contractual workers. Duty regarding coordination between the primary contractual workers will lay on the project group, as picked by the proprietor and headed by the project supervisor.

Consumption increments quickly amid the Construction work due to the elevated amounts of labor work, plant and materials required in the work. Any disturbance to the productive utilization of these assets will bring about huge cost invade. Severe harm should be possible to the project program and spending plan by change, deferral and disturbance. Nonetheless, the project administrator must be careful that endeavors to expand advance or decrease use don't risk safe working practices or the nature of work performed.

The project administrator will depend on the project group to keep him evaluated of advances and costs and will examine week after week or month to month reports, hold consistent advancement meetings so as to determine current challenges and foresee and resolve set strategic and legally binding issues. The project administrator will likewise wish to guarantee that adequate records are being kept up by the group to protect the proprietor's legally binding position and to fulfill the prerequisites of the statutory and certifying authorities and the possible administrators of the plant. (Tonchia, 2008)

The project administrator should convey that advancement reports don't just report advancements accomplished, however that they represent the exercises that have not happened, since it is these "negative" parts of the report that give early cautioning of future issues.

Finish of construction work regularly implies that the contractual worker has raised the plant as per drawings and determinations, finished any predetermined pre-start up work and finished the last tidy up. Be that as it may, as-fabricated drawings, operations manuals and handover of other indicated records will be required before the full contract is worth finish. (The Joint Development Board sponsored by the Association of Cost Engineers and the Royal Institute of Chartered Surveyors, n.d.)

INITIAL OPERATIONS

Introductory / initial operations is a term used to portray the whole procedure of pre-commissioning, commissioning, beginning start-up and production and execution testing of a plant.

Different terms are utilized by various organizations to cover the exercises required to tweak a plant and bring it into production and the terms and definitions utilized here are characteristic as it were. The arrangement of exercises driving to start up and acknowledgment of a procedure plant will commonly be as takes after:

Construction

⇓

Pre-commissioning

⇓

Mechanical completion

⇓

Commissioning

⇓

Mock operation

⇓

Initial start-up

⇓

Check product specification

⇓

Check production performance

⇓

Acceptance of plant

Pre-commissioning and mechanical completion

Pre-commissioning exercises are those which must be attempted earlier to working of the equipment, for example, alterations, cold arrangement inspection of machinery so forth performed by the construction contractual worker preceding commissioning and without which the establishment and installation of the equipment can't be said to be mechanically entire.

(Adedeji Badiru, n.d.) Mechanical completion of a plant or any piece of a plant happens when the plant or a piece of the plant has been finished as per the drawings and particulars and the pre-commissioning exercises have been finished, to the degree important to allow the proprietor to acknowledge the plant and start commissioning exercises. Mechanical completion will ordinarily be considered to have been accomplished on fulfillment of the accompanying:

- Hydrostatic testing of all pipework and equipment
- Testing of electrical frameworks and specialist equipment bundles
- Cold check of machinery and equipment
- Circle checking of instruments
- Flushing of pipework
- Acknowledgment of all documentation as per the methodology,
- Drawings and specs

Commissioning

A plant, or any characterized piece of a plant, is prepared for commissioning when the plant has accomplished mechanical completion.

Commissioning exercises are those related with planning or working the plant or any piece of the plant preceding the underlying start-up and are much of the time done by the proprietor or a joint proprietor/contractual worker group. Commissioning may include ridicule operations which are startup exercises leading to permit operational testing of the equipment and operator training and acquaintance.

Toward the completion of commissioning, the plant will be completely prepared for production operation.

Initial start-up

Beginning start-up happens when feedstock are acquainted with the plant for the express reason for delivering a saleable product interestingly, frequently alluded to as 'oil-in' in the hydrocarbon preparing industry.

PROJECT RISK ANALYSIS AND MANAGEMENT

Risk analysis

Outrageous happenings, for example, force majeure, can't be disregarded when settling on a choice to endorse a project. In spite of the fact that on a clear project such dangers can be considered without expound estimation, on a noteworthy project the level of instability associated with cost, time, area, innovation and quality and the intricate way in which the different dangers collaborate to amplify or quench each other, must be completely investigated, substantiated and introduced by undertaking a far reaching investigation of the full scope of dangers included. The very procedure of undertaking the examination gives huge help in limiting danger since it centers the project group on those territories requiring most noteworthy consideration. (George A. Steiner, n.d.) Risk analysis is a device that gives a formalized way to deal with the ID and assessment of an intricate arrangement of vulnerabilities, their probabilities and interdependencies and gives a rationale and center to the proceeding with administration of risk. Since Risk analysis incorporates into its appraisal the likelihood of the event happening and exhibits this data in a way from which the leaders can attempt their cost and advantage investigation, it is fit for dealing with arrangements against all sorts of hazard, including force majeure and other outrageous inevitabilities that would not regularly be incorporated into a customary contingency allowance.

The expanding utilization of PCs has supported the advancement of refined methods for the investigation and administration of risk on construction and infrastructure development projects. While these strategies are of undoubted incentive to a project, their utilization must be expected at initiation since it expects revisions to the way in which decisions are set aside a few minutes required to embrace the examination must be permitted in the program set up for the project.

Likewise with a customarily figured contingency allowance, the arrangement for risk can be an extremely noteworthy sum inside an estimate and not just should be taken in a way proper to a noteworthy cost component, however additionally requires the examiner to have the capacity to clarify the premise on which risk was computed and the likelihood of likely event of the event. (Adedeji Badiru, n.d.)

Identification of risk

The initial phase in solution of an issue is realizing that the issue exists and in risk administration the most essential stage is the distinguishing proof of territories of potential hazard. (The Joint Development Board sponsored by the Association of Cost Engineers and the Royal Institute of Chartered Surveyors, n.d.)The different dangers can at first be distinguished by:

- Reference to an agenda;
- Subjecting the project group to a meeting to generate new ideas;
- Grouping records arranged by people or divisions;
- The investigator from individual experience

Evaluation techniques

There are three sorts of risk investigation in like manner utilized:

- Elementary
- Sensitivity
- Probability analysis.

Each kind of investigation offers contrasting degrees of modernity and each compares to the specific necessities of the project at the season of examination.

Elementary risk analysis

It is a Basic risk examination and it depends on the judgment and experience of those undertaking the investigation and essentially contains the assertions of the impact of the most clear real dangers expected on a project. Basic risk examination is ordinarily utilized on an uncomplicated project or at the examination phase of a noteworthy project.

Sensitivity analysis

Sensitivity examination considers the effect that fluctuating degrees of a solitary component of risk may have on the project, by breaking down and assessing its impact on the program or spending presumptions.

A Sensitivity examination is every now and again exhibited graphically, or as an arrangement of figures, yet its shortcoming is that it treats every component of hazard independently and does not focus on the unpredictable effect on parallel, or later, exercises. Sensitivity investigation is oftentimes utilized later in the project life while assessing tenders and considering the effect on cost of design or time changes, and so forth.

Probability analysis

A probability analysis will assess the numerous components of risk inside a project and will allot a level of likelihood and scope of qualities to each. Thusly an evaluation can be made of the most likely result, with measurable spread of different outcomes. In its most refined frame a probability analysis tends to the important effect on the continuous exercises and dangers.

In spite of the fact that a summation of the most likely outcomes can be made, it is far-fetched that practically speaking the most plausible outcome will be accomplished against every component of hazard; who is to state that they are the most plausible outcomes? To a degree it is a lottery which risk happens and to what degree. The power of the PC may in this way be utilized to produce an extensive number of arbitrary situations, in view of the surveyed likelihood factors, to give a no doubt last outcome. It is also alluded to as a Monte Carlo simulation. (George A. Steiner, n.d.)

Assessing probabilities and values

To give the fundamental data to the examination, design engineers, cost engineers, development directors and others encountered in their controls might be made a request to acknowledge or alter a rundown of conceivable dangers, to express a base suspicion and the probability in rate terms of accomplishing that base. The risk examiner will then talk about with the specialist the most material state of dispersion covering the spread of conceivable outcomes some time recently encouraging the data into the PC. The hazard investigator will utilize the planning system to consider the effect that the different classes of hazard will each have on the other, while taking consideration not to copy chance by counting an individual hazard that is somewhat secured by different dangers. A case is the danger of time invade in the creation of a piece of the work, which might be distinguished in that capacity in the investigation; yet part of this risk might be because of postponement in conveyance of materials, which might be an independently distinguished risk thing.

A probabilistic risk investigation will create values and related certainty levels on a "S" bend, running from the assurance that the project can't be finished beneath an expressed figure and aggregate certainty that it will be finished inside a further figure. In the mid-scope of the "S" bend there will be a scope of projected last worths, each with a related certainty level. The choice to continue with a project, or its next stage, might be taken at a financial plan assume that has a certainty level of half, i.e. where there is an equivalent possibility of under running or overwhelming the figure. On the other hand, relying upon the normal return or other advantage that the project will give, the choice might be taken to authorize the project at a lower spending plan, yet with a low level of certainty that this low figure can be accomplished, or the other way around. (Adedeji Badiru, n.d.)

A full probabilistic risk examination can consequently give leaders with a more noteworthy comprehension of the conceivable aftereffect of the choices taken and methods for judging more precisely the more serious risk that might be satisfactory on a lucrative project.

Risk management

One of the fundamental obligations of the project group is the administration of risk. Risk can be overseen by transference of risk to others, by shirking of risk, or by controlling the elements causing the risk.

Transfer of risk

Risk can be exchanged by setting the duty regarding its result on others, e.g. by using protection or by asking for settled costs instead of paying for expanded expenses as they happen. Risk is in this manner for the most part exchanged at a cost.

Avoidance of risk

Where risk is because of a factor that can be changed the open door exists to stay away from the risk by changing the factor, e.g. by changing from new untried innovation to an attempted and put stock in elective, or moving a touchy operation from winter to summer. Risk is in this way typically maintained a strategic distance from by tolerating what might be viewed as a less alluring project choice.

Control of risk

The initial move towards the control of risk is the distinguishing proof of the major risk ranges and their underlying assessment took after by proceeding with control. This control is practiced through the ordinary project control strategies of consistent advance estimation and correlation against a characterized base, the distinguishing proof and execution of required remedial activities and the checking and detailing of aftereffects of the move made.

Risk register

Once distinguished the different risks will be consolidated into a risk register which regularly contains the accompanying data against each risk thing:

- The possible effect of the risk as far as cost
- A concise articulation with respect to the foreseen methods for relieving the risk;
- The name of the individual or office in charge of alleviating the risk

As the project grows, new risks will be recognized and before risks will change in status. The project group will in this manner be welcomed occasionally to re-assess old risks and to recognize new ones for incorporating into the register.

Once the risk register is issued the mindful individual will utilize it to organize his consideration on those risk things which should be put into concentrated care with a specific end goal to keep their belongings to a base.

SAFETY, ENVIRONMENT AND QUALITY ASSURANCE

Safety, environmental and quality prerequisites should be built up and arranged ahead of schedule in the life of a project. Safety is critical both in construction and operation. Safety must be arranged, as far as sheltered design and in the selection of contractual workers who are equipped for working securely.

Environmental assurance policies are under steady audit, which makes forecast of necessities around this area a great deal more troublesome and much more project particular.

Extensive consideration is in this way important not exclusively to get the safety and environmental issues right, however to agree to the changing administrative regime.

It is of prime importance that a manager who is educated of the enactment is made in charge of these issues on a project. Such a part may be well advantageously incorporated by the quality affirmation part as they are firmly interrelated.

Safety plan

The system important to meet the necessities of the safety controls will be epitomized in a safety plan for that drives the advancement of the major safety exercises. The safety design has the reason for specifying the major safety exercises (studies and obligations) at each phase of project advancement. As each new stage is achieved the safety issues turn out to be more particular, the safety plan is created with progress of the project. (The Joint Development Board sponsored by the Association of Cost Engineers and the Royal Institute of Chartered Surveyors, n.d.) The targets of the safety plan are:

- To demonstrate the real safety exercises
- To characterize team duties
- To start studies and screen the work
- To evaluate examinations and spread the conclusions, suggestions and activities.

Safety philosophy

The safety philosophy is an high level article and is the beginning stage in depicting how safe design, construction, operations and support will be accomplished.

The targets of the safety philosophy are:

- To give clear explanations and heading to the improvement of 'safety prerequisites'
- To guarantee that safety is joined into all design work
- To start positive consideration of danger disposal
- On account of seaward establishments to start the arrangements for clearing departure and reuse, and transitory safe asylum

The philosophy ought to be various leveled in approach and ought to look at safety to individuals from the worldwide issues through position, introduction, design, equipment and safety frameworks.

Safety requirements

The safety prerequisites are built up by adding point of interest to the safety philosophy and will be a document that will be utilized by the design contractual workers, furthermore, the project group to screen safety design work. The goals of the design safety prerequisites are:

- To give a report which in clear unmistakable terms points the safety necessities for the planned establishment and installation
- To give a springboard from which engineers of various genres can assume liability for the safety necessities particular to them also, to form them into the design.

Environmental impact assessment

The environmental impact assessment, where required, gives administration and statutory bodies with a formalized and institutionalized methods for guaranteeing that the established plant performs inside administrative and corporate environmental measures. It evaluates the aggregate environmental effect of potential releases, counting the level of risk and potential seriousness of inadvertent arrival of toxins. It additionally addresses the pertinence, practicability and adequacy of plant and equipment, strategies and work force in the aversion of environmental harm and the moderation of the impacts of all releases regardless of whether inadvertent or operational. (George A. Steiner, n.d.)

Quality assurance

Quality management applies to all products and services and all administrations and a certified quality affirmation/quality control framework is getting noticeably more important with the end goal to exhibit that an organization is equipped for accomplishing a reliably high caliber of work.

The target of the quality control is to give a framework that bolsters the quality confirmation get ready for the project, and guarantees consistence with controls covering the establishment of a facility.

Certification requirements

The certification prerequisites for inland and seaward projects have much variety. The project ought to build up a project particular confirmation and administrative prerequisites plan posting every one of the necessities, together with points of interest for the fitting administrative body or delegate and lead times for endorsement entries.

INSURANCE

The sorts of insurance required on a noteworthy engineering project are regularly exceptionally broad and the scopes, funds protected and points of confinement of reimbursement must be deliberately evaluated, requiring an unpredictable strategy proclamation. A normal engineering development project is mind boggling, of high worth and requires the arrangement of more than one noteworthy contractual worker. It is in this way regularly the case that the proprietor will determine in the agreement conditions that he will organize certain predefined insurances. (Tonchia, 2008)

Such insurances are probably going to cover as a base the arrangement of an all risks arrangement on the changeless works, including material misfortune or harm to the work attempted and in the meantime ensuring the interests of the proprietor and his contractual workers. In specific cases, and particularly where work is being done in and around existing plant of high worth, the proprietor may likewise acknowledge duty regarding orchestrating outsider insurance.

Project insurance

A proprietor's all risks policy will cover the perpetual works, materials for construction of that and will once in a while reach out to transitory works, constructional plant and equipment. The cover will apply on the site of the works, material and equipment in travel and maybe to off-site work. Either in a similar policy, or by a different insurance, the 'all-risks' might be reached out to incorporate a repayment against lawful obligation for damage to third party and harm to their property for a particular cutoff of repayment. (George A. Steiner, n.d.)

Contractual worker's insurance

The degree to which a contractual worker is in charge of misfortune or harm to the works and material for construction will be determined in the states of agreement. Moreover a contractual worker will ordinarily be capable (once in a while without response to the proprietor even where careless) for loss of or harm to constructional plant and equipment and temporary settlement. The contractual worker is required to embrace by law certain insurances such as boss obligation in connection to his representatives and insurance of engine vehicles on general society interstate. Other insurance planned will rely upon the legally binding commitments and generally be founded on the philosophy of the contractual worker's administration.

Claims and reporting

It is typical for the agreement conditions to require a contractual worker to report all misfortune or harm quickly to the proprietor, there being from there on a laid down method for the declaring of claims and the settlement of misfortunes contingent upon which party has organized the insurance.

CHAPTER 3

Estimation and Assessment

INTRODUCTION

The capacity to estimate cost precisely is vital in all enterprises, however necessary where large amount of investment is to be focused on a venture whose last cost is not be known from 10 years or more. A poor estimate can in such conditions empower interest in uneconomic or projects with negligible profits while leaving other projects kept from cash, and may cause the losses for an organization. On the other hand, project opportunities might be missed with the conceivable danger of a contender putting resources into that project.

In the engineering enterprises the proprietor will try to estimate demand and subsequently the estimate the worth of a product, adjust this against the cost of production, and ensures a satisfactory profit for the capital contributed. The cost of production will itself include estimates of crude materials, the capital and financing costs, and the operational expenses included in tailoring of the crude materials into the product, including market dissemination of the product and sales. (Salvendy, n.d.) This chapter describes the estimation of the capital cost of production set ups up to the planning of the most itemized estimate, regularly called the complete estimate or definitive estimate.

Initial actions

On receipt of a demand for an estimate, the estimator must make immediate contact with all the key staff associated with the project to decide:

- The program for finishing the estimate which should cover plan arrangement of drawings, specifications and like things
- The Nature of the plant;
- The nature of accessible data
- The scope of design plans accessible;
- A satisfactory base case;
- The date for access to extra information and data
- The precision and timing of the estimate which is required.

The estimator should acquire data with respect to:

- The kind of local environment
- Availability of local facilities and offices;
- The position of the plant with respect to different developments and whether such different developments are "live" plants;
- Relationship of the plant to associated establishments;
- Whether associated establishments give facilities and support to the new plant or are reliant on it.

The estimator will subsequently not just consider work inside the assigned range of the plant, additionally the way in which the plant coordinates with other establishments that might be giving the power, water, putting out fires systems, etc. required for benefit of the plant. Such systems are, in a petrochemical environment, usually alluded to as utilities. Additionally, thought is given to the pipelines or different methods for conveyance, stockpiling or expulsion of the crude materials and product (regularly alluded to as offsets).

Once an understanding is made of the idea of the project, thought is given to the data promptly available to hand, which can be utilized specifically or in an adjusting manner in the estimate. This data might be in house or distributed, yet it will be of little utilize and emphatically perilous if points of interest are not accessible that place the different expenses into setting, e.g. location of area, date of development and construction, capacity limit, feedstock, nature of the ground and arrangement of utilities. (Salvendy, n.d.)

Things much of the time underestimated or overlooked from an estimate include:

- Utility supplies
- Extent of access to stages
- Temporary laydown areas for furnishing work before erection;
- Foundations for substantial lift cranes;
- Spares and appointing spares;
- Initial charges or substances for the equipment;
- Logistical prerequisites in getting substantial quantities of men and vast amounts of equipment on to a site.

Other variables can encompass:

- Skills which are required and not locally accessible;
- Local convenience not satisfactory to house the workforce;
- Bridges not adequately solid to convey the vehicle loads;
- Electrical cables too low or curves in streets excessively intense, making it impossible to take long transporters.

Nothing unless there are other options can be expected from a plant Layout, and it is in this manner fundamental that the estimator ought to assess the site, the nearby environment, the fundamental get to courses and view the construction techniques as used to assess any such subordinate cost. (The Joint Development Board sponsored by the Association of Cost Engineers and the Royal Institute of Chartered Surveyors, n.d.)

Inland and seaward estimation agenda checklists are accessible from the Association of Cost Engineers.

Estimating accuracy

An estimate is basically reliant on four factors, each of which will have a critical effect on the exactness of the figures delivered:

- Availability of outline;
- Availability of pertinent cost information;

- The ability and experience of the estimator;
- Time frame which is set to deliver the estimate.

At the point when every one of the four of the above are promptly accessible, a precise estimate ought to be conceivable, yet in the early phases of a project, design and outline would not be accessible and dependence will be put on recorded cost information and involvement in the utilization of such information. Interesting projects will require extra cost research to insert the accessible verifiable information, which additionally emphasizes the significance of an accomplished estimator's capacity to translate the simplest idea into a valuable estimate. It will be essential at the soonest stages to qualify the estimate with respect to any assumptions made which will significantly affect the estimate. (Horngren, n.d.)

As a project builds up, the level of detail inside the estimate, and along these lines the precision of the estimate, increases. Not exclusively does the technical definition increase amid the length of the project, additionally the number of design plans decreases, along these lines diminishing the level of plan instability and changeability. (Salvendy, n.d.)In spite of the fact that a high level of exactness is always needed, there is no justify in endeavoring to accomplish a level of precision wrong to the data accessible and improper to the choice that is needy on the estimate. An estimate created from least data can't be obviously enhanced by endeavoring to include disconnected things of detail, and in addition may give it an apparent validity that it doesn't merit.

The estimator will be required to deliver an announcement with respect to the exactness of the estimate comprehensive of a contingency allowance as talked about later in this chapter. This announcement is ordinarily communicated as add or subtract range, which is commonly alluded to as tolerance limits, which will change subject to the sort or class of estimate. The points of confinement of the range are the least and most extreme estimations of the estimate and are communicated as a rate of the normal cost. Since there is a farthest point on most reduced conceivable cost, i.e. zero, however no identical maximum point of confinement, this 'certainty factor' will regularly have a positive figure higher than the negative. (Salvendy, n.d.) The precision proclamation must be self-assertive, and each estimate ought to be assessed all alone with its merits. Nonetheless, since the judgment of assessing precision is fundamentally based on the nature of data accessible, including extension, cost information and productivity, an accomplished estimator ought to be ready to recognize the hazard and possibility prerequisites and forecast a sensible precision level.

STAGES IN ESTIMATE PREPARATION

A project progresses by making measured strides through a minefield of checks, cheques and parities, including recommendations, estimates, endorsements, audits, changes and further approvals. Every endorsement and approval will be judged against prior endorsements and the proprietor of the firm will search for the purpose behind any change, particularly if such is critical to the practicality of the project.

Estimates will be required on a project from the date on which it shows up as a splendid idea in the creative mind of a firm's proprietor, to the date on which the project is finished and given over. The number and sorts of estimates to be attempted in the term of the project will rely upon its multifaceted nature and complications.

The production of an estimate is itself a not immaterial cost, and stress might be there to diminish the quantity of estimates or to restrain their degree. In such conditions, there will be a need to adjust the cost of setting up an estimate against the hazard in not setting up the estimate. The primary phase of a project (the plausibility stage), will start with the requirement for an operating facility being set up. The definition gave to an estimator as of now will be constrained to a portrayal of what the facility needs to accomplish to meet the proprietor's prerequisites. The subsequent estimate will be at its most surmised, however since early estimates are utilized exclusively to choose to focus on a moderately humble extra interest in doing further design, the level of exactness required may be with respect to the speculation choice being taken, e.g. regardless of whether the project is probably going to be suitable also, henceforth the design worth creating. (Salvendy, n.d.)

Following ending of the feasibility stage and acknowledgment of the estimate, a restricted measure of design and development work will be attempted to create what may at present be sketchy recommendations. This is typically alluded to as the conceptual design stage and will cover the exercises recognized in the Table. Conceptual design might be attempted by the proprietor or the proprietor's picked planner yet will dependably be made by a firm's most experienced work force. This will be the primary endeavor to depict and focus on paper the extent of a plant fit for meeting the stipulated prerequisites, the attributes and yield of the plant and will distinguish the required processes and real equipment things together with a conceivable design. (Horngren, n.d.)

The cost of point by point design and configuration is generous and can include substantial designing groups. Time spent in the basic parts of proposing and marking down different design and outline options inside a specific idea can be expensive in both time and hours. It is along these lines habitually the case that a further stage is presented between the conceptual and point by point, detailed design stages. This further stage might be utilized to add facilitate definition to the theoretical design and is usually named preparatory design. On the other hand, it might be utilized to approve a few parts of a very much created conceptual design or as a first stage of detailed point by point design, named front end design.

The choice to embrace either preparatory design or front end design will be depending not just on the status and trust in the conceptual design additionally on the typical methods of the specific organization or project supervisor.

ESTIMATE TYPES

As already expressed, the sorts of estimate are numerous, they have fluctuating profundities of value and are suited to very unique and unmistakable purposes.

Each estimate is expected to anticipate the last cost of the design and execution of a project, and each is fit for a level of complexity and precision fitting to the phase of the project.

It has been recommended that the early estimates are not completed by the evaluation of parts but instead by the utilization of recorded, all inclusive rates to product criteria.

Estimate substance and extents will obviously shift between contrasting sorts of project, and will change amid the life of a project, reflecting not just the level of detail accessible additionally the motivation behind the estimate.

Table 3 1: Estimate exactness (The Joint Development Board sponsored by the Association of Cost Engineers and the Royal Institute of Chartered Surveyors, n.d.)

Technical definition	Project stage	Estimate type	Estimate accuracy (%)
Generalized scope definition			
Required capacity of plant			
Plant location			
Preliminary project schedule			–25 to + 50
Preliminary process diagram			
Main statutory requirements			
Equipment list			
Outline engineering specifications			
Preliminary 'block' plot plan			
Offsites and utilities by system and capacity			–15 to + 25
List of major equipment			
Preliminary equipment data sheets			
preliminary process and instrument diagrams			
Quotations for unfamiliar items			
Detailed scope definition			
Finalized heat and mass balance calculations			
P & I Ds[a] for process and offsites			
Detailed plot plan			
Detailed engineering specifications			
Operator's requirements			–10 to + 25
Local authority requirements			
Project master schedules			
Information on site conditions			
Local availability of labour and material			
Detailed equipment list			
Completed and approved plant layout			
Electrical single line diagrams			
Detailed equipment specifications/data sheets			
Detailed material take-offs			
Firm quotations from potential vendors			
Quotations from potential contractors			
Commissioning and operating infomation			–5 to + 10
Installation and fabrication specifications			
Amended design and specifications a a result of safety reviews			
Construction subcontract enquires			
Production design phase continues			

Process and instrumentation diagrams

Key to design stages: Feasibility

Conceptual design

Preliminary front end design

Detail design

Key to estimates types: Order of magnitude

Appropriation

Budget

Definitive

An estimate produced in a formalized way, with an eye to its proceeding with utilize as a wellspring of information and control amid the project and as a receiver of data that adds to the in-house information base, will have various appearances. Every indication ought to preferably be an advancement of that which went earlier and ought to oblige what is to come.

The exemption to the above might be the order of extent estimate, which is every now and again on a premise not quite the same as resulting estimates. It might be founded on the ordinary cost of a plant equipped for handling an expressed measure of product every day, which as an arrangement of criteria is inconsistent with ensuing estimates, depending as they do on the distinguishing labor and materials required to design and build the plant, as opposed to the capacity limit of the plant.

In any case, every consequent estimate have a progressive relationship, each estimate containing a more prominent level of detail inside a static arrangement of accounts that can be identified with the delicate enquiry records and broadened into project execution to give a premise to cost control. Great cost control will keep the last estimate live, with old data being supplanted by new for later investigation to give the high level information to future order of extent estimates. Thusly a reason for examination exists all through the project life and information are given to use in following projects. (Salvendy, n.d.)

There are four essential sorts of estimate that are for the most part required amid the different phases of the project, each of which might be upheld by various supplementary estimates to evaluate middle change and worth alternatives:

- The primary estimate amid the feasibility stage, differently alluded to as a conceptual definition, class IV or order of extent / magnitude estimate;
- The second estimate preceding sense of duty regarding the further cost of conceptual design, differently alluded to as a blueprint definition, class 111,study or assignment estimate;
- The third estimate required amid the front end design stage preceding sense of duty regarding the critical cost of definite design, differently alluded to as a design, class 11, preparatory or spending estimate;

- The fourth estimate before beginning of capital consumption, differently alluded to as an execution, class I or complete estimate.

Estimate titles will depend on the business as well as on whether the estimate is created by a proprietor or contractual worker. Each organization will have a favored title, class definition and demonstrative precision ranges, however estimate sorts are alluded to in this as:

- Order of magnitude / extent estimates
- Appropriation or allotment estimates
- Budget and spending estimate;
- Definitive or Complete estimate.

Moreover, there might be extra middle of the road stages, or on the other hand, a converging into less estimate sorts to suit the individual needs of particular organizations. It is in this manner prudent to express the plausible precision, together with the estimate portrayal, at whatever point such estimates are offered or asked for outside the organization / firm.

Table shows the periods of the project design amid which the specialized definition is created and the level of exactness ordinarily expected of the four essential estimate sorts.

The estimator will focus on the basic cost regions, remembering the further impacts that such basic things may have on different components. A normal funneling account inside a spending estimate for a petrochemical project would be viewed as basic, since it might liken to 20 to 25% of aggregate material costs, 40 to 45 % of aggregate work expenses and besides will have an effect on pipe bolsters, foundations, painting, protection, engineering and project administration costs. Channeling accounts on an engineering project additionally have the most prominent inconstancy and are the most hard to estimate. (Salvendy, n.d.)

By differentiate, the steel structural work component of a run of the mill seaward module would be viewed as essential yet not basic since despite the fact that it might constitute a comparable extent of the material and man-hour cost as the past case, it doesn't have an indistinguishable effect on different exchanges from it is designed right on time, in a conventional way, and does not for the most part change drastically.

Order of Extent / magnitude estimate

Some of the time known as a "ballpark" or 'seat of the jeans' estimate, a request of extent / magnitude estimate is delivered for the quick assessment of business potential outcomes and monetary outcomes of a project. It is oftentimes in light of net engineering parameters.

Despite title, this kind of estimate will be utilized to increase beginning approval for improvement kinds to proceed into the following phase of the project and will contain (moderately) less detail. It will be reliant on historical information and will utilize comprehensively based all- inclusive prices and rates. A request of extent / magnitude estimate will ordinarily be considered to have an exactness of - 25 to +50%.

Since minimal point by point / detailed information of a plant will exist at such an early stage, the estimate will rely upon star rating data from past projects that contain comparative processes, things, structures, frameworks and components beforehand designed, fabricated and priced. Modifications will be made for time, area, changes in economic situations, current design prerequisites furthermore, relative capacity limit before their utilization in the estimate.

Cost does not shift directly proportional to the size, and real changes in extension or relative capacity limit between the present project and the project being utilized as an evaluating base, can be balanced utilizing a factor here and there alluded to as the 'six tenths' rule, this being a simple proportion by which cost expands with respect to size measure. This dependable guideline expresses that the sum by which the estimated cost of a coveted thing ought to be balanced in respect to a change in size or capacity limit, can be ascertained as the pro rata change multiplying by six-tenths. On this premise the estimated cost of a thing of plant that pairs in size will increment by 60%.

The reexamined composite rates acquired by such modification, or molding, give a sensibly precise and quick estimate as well as will likewise be utilized to give a premise to cost control as the design advances. Obviously, the level of exactness accomplished from such a "calculated" estimate will rely upon the accessibility of verifiable information from comparative projects in terms of scope, area and scale, feedstock, economic situations, and so forth. Information acquired from past estimates may have begun either from tenders got, or from conclusive expenses brought about, and for each situation may need to be balanced further, either on the grounds that an offered figure will perpetually be less than the last cost, or then again, a last cost may incorporate the cost of overcoming issues that may not repeat.

On the off chance that verifiable information from a past project which is not a close copy of the proposed project are to be utilized, additionally innovative work will be required to accomplish the level of exactness required, before responsibility of development is funded.

Without data from a close copy plant, the estimator will depend on distributed or recorded data from various existing projects to follow the difference in cost of the foremost segment parts emerging from contrasting obligations, limits, connections to existing equipment, or, on the other hand the idea of the feedstock.

Recorded data is of little use without an information of any capability that made such information one of a kind in entire or part, i.e. nature of ground conditions, control era being given by existing installation, liberal or congested plot etc.

Appropriation / allotment estimate

The Appropriation / allotment estimate is once in a while known as a class 111 estimate, since it utilizes specialized data created to a level of definition portrayed as class 111.

In spite of the fact that there is as yet an absence of design detail important to accomplish even a preparatory measurement of the components included in the project, the designers will now have distinguished the significant equipment things and decided their required yields. This data ought to be adequate to permit a more precise estimate to be made of the cost of the real things of equipment, and will give a chance to make enquiries of potential providers in regards to key parts including spending cost. Such key segments can include a noteworthy extent of the aggregate capital cost, also, the catch of these essential expenses with a level of assurance will give added certainty to the estimate. (Horngren, n.d.)

The rest of the components will even now not be characterized to an adequate degree of detail to enable vendors to give spending costs, or to permit a measurement adequate to attempt whatever other than a factorized approach, also, without such better data, the cost of establishment and installation of equipment can be estimated by a Lang factor. The Lang factor is termed after the name of H.J. Lang, who found that different sorts of plant had a sensibly steady connection between equipment cost and installation and establishment cost. (Salvendy, n.d.)

An Appropriation / allotment estimate will along these lines be constructed considerably in light of the rundown of significant equipment. Expenses will be estimated utilizing as a part of house information or preparatory spending costs, acquired from vendors for its production and installation, however with a large portion of the rest of the construction components included as a rate of the estimation of everything of equipment. Along with a decent information base, such strategies can create an extremely worthy outcome, inasmuch as the nature and premise of the information are completely understood.

An Appropriation / allotment estimate will usually be considered to have an exactness of - 15% to +25%.

Vendors will every now and again recommend elective alternatives and thought ought to be given to such extra data to streamline costs.

Budget estimate

Generally called a class I1 estimate, the Budget and spending estimate will be delivered once the calculated designing stage has been finished. It will be utilized as a part of the official conclusion to resolve to detail design and to give a premise to the proceeding with cost control of the project. The spending estimate will regularly be considered to have a precision of - 10% to +15%.

In the early phases of a project, design hours for engineering, acquirement, project services and project administrations will be incorporated into the order of size and evaluation estimates, either by consideration in a combined rate or used as a percent rate. In the later phases of Budget and spending plan and authoritative estimates, these same hours will ordinarily be independently estimated by the individual client offices. This will as a rule be finished by the design, who will envision the quantity of drawings or demands (expectations) required, or on account of administration, by thought of the association required to attempt the work, and the term of each of the positions recognized on the association diagram. (Horngren, n.d.)

Data accessible at this stage will take into account estimated amounts of control costs acquired from potential vendors and contractual workers. Such amounts, albeit preparatory, will regularly be accessible from material departures created from drawings, either physically or electronically.

The cost of construction work labor for incorporation in a financial plan or authoritative estimate will more often be computed by applying man-hour standards to the measured calendars and appending comprehensive work rates to the subsequent sums. The labor rates to be utilized will incorporate the cost of "ordinary" construction plant, supervision, site convenience, and so forth however will reject major technique related construction plant, for example, substantial lift cranes and so forth.

The material and equipment components inside a financial plan or authoritative estimate will be ascertained or refined by the issue of preparatory enquiries for spending costs, once the vital data is accessible. On account of the real equipment things, such enquiries may not be preparatory but rather Or maybe the demand for affirmation of past estimates, since the vendors will have been made a request to give preparatory data identifying with execution, capacity limit, different parts of design and estimated cost right on time in the conceptual stage.

Definitive or Complete estimate

The last estimate delivered quickly following sense of duty regarding the major capital use, is the Definitive or Complete or class I estimate and will typically be thought to be to an exactness of - 5% to +lo%, along these lines giving little edge for mistake.

This estimate will ordinarily contain the level of detail utilized as a part of the execution of projects and the planning of bids. It will be utilized as a part of the support of close control over the cost of the work, or for designation of assets into work packs, work cards and comparative project control instruments.

The Definitive or Complete estimate will utilize a lot of the data gotten amid the budget estimating stage, however will have the advantage of more Nitti gritty design particulars, detailed labor work and material breakdowns together with more precise costs from responsibilities made, or enquiries issued to vendors and potential temporarily hired workers - contractual workers. (Horngren, n.d.)

While the Definitive or Complete estimate is more precise with respect to amount, detail and cost, since the estimate is being delivered at a high level stage utilizing particular detail, it is more defenseless to blunder by exclusion than is, for case, an order of magnitude estimate. The order of magnitude estimate will most likely have been founded on an examination of the real final cost of a past project, a 'top down' estimate. By it's exceptionally nature, along these lines, an order of magnitude estimate incorporates all expenses acquired on the past project. Interestingly, a complete estimate is a 'bottom up' estimate, developed from data particular to the project and which can't depend on a definitive security net of a comprehensive rate, to represent all the sundry things that should be considered when arranging a noteworthy complete estimate. Such sundry things, if not independently tended to, might be incorporated as a rate add on to the estimate, or on the other hand, included inside the rates furthermore, costs utilized.

Different on-and seaward checklist agendas are distributed to help the estimator furthermore, help avoiding from oversights from the estimate

HIERARCHICAL COMPOSITE RATES

All estimates are the consequence of an amount being multiplied by a rate, all the rates other than the most essential expenses are made out of various cost components in different proportions to shape a composite rate. Composite rates may reflect detail at different progressive levels; the largest amount of detail might be the fundamental expenses of labor, plant, materials, and so on. while the most minimal level could be a solitary comprehensive composite rate per unit of production of a finished plant.

In spite of the fact that the utilization of low level composite rates in an estimate can be both financially savvy and exact (if legitimately delivered and recorded), the utilization of such rates without a comprehension of their original substance is risky. A full arrangement of composite rates at different levels, with no thought of what the rate speaks to, when the rate was present, and where they were utilized, is of little advantage to any organization and it is hence prudent to keep up a record of the essential parts of composite rates, including design criteria, date and location of area, with the goal that rates can be adjusted for use in various conditions. (Salvendy, n.d.)

Not at all like the building and structural engineering enterprises that tend to estimate utilizing composite rates that are comprehensive of labor, construction plant and materials. The engineering enterprises running various chemical processes estimate each of these components independently. The building and structural engineering segment of a industrial engineering project may at first be created utilizing strategies suitable to those enterprises. But, it is not irregular for the estimator to break down the coming about estimate into its constituent parts for steady introduction and to anticipate more precisely general work levels required on site and to foresee the need to legitimize the work stacking inside what is thought to be a sensible aggregate workforce.

The rate for labor utilized by an estimator will be a type of composite rate generally alluded to as an 'all-up' comprehensive rate. An 'all-up rate' can be translated as including a few or every one of the expenses of employment, arrangement of convenience, administration costs, development plant, material expenses and the like. It might have been delivered from the examination of the aggregate finance alone, the finance for a specific genre, or got by partitioning the last outturn expenses of a project by the aggregate nearby hours spent. An estimator should in this manner dependably qualify his demand for data, or his reaction to a comparative demand, by expressing its considerations by and large terms.

This proceeding issue of definition can be additionally confused when a rate is expressed as far as immediate worker hours, as the estimator must be cognizant to the further issue of definition, as examined beneath.

The message all through this text is that any rate that is not characterized is unusable, and an accomplished estimator when offered a rate will dependably react with the inquiry 'what does that mean or incorporate?

MAN-HOUR NORMS / STANDARDS

In the engineering ventures, rather than the building and structural engineering businesses, there is an inclination to estimate work and material independently. Engineering plants are as often as possible developed in remote regions or cities and work productivity and proficiency is thusly an issue of particular thought. Most organizations have an arrangement of "base" man-hour standards; such standards may have been initiated by themselves, or other than this the organization may use as a base a set of published and distributed standards. published / Distributed standards will normally be at the most elevated level at which work can be measured, i.e. per meter of a predefined pipe or, on the other hand per butt weld, and the estimator may require to assemble information on composite rate similarly as has been talked about for composite rates. (Horngren, n.d.)Whatever their starting point, standards will mirror an expressed arrangement of conditions what's more, will expect acclimation to accord with the levels of proficiency anticipated of the new plant by the use of a proficiency factor. The proficiency factor to be connected to such essential standards will consider not just of the estimator's involvement of his organization's current execution against the built up standards, additionally the type and location of area of the work for which the estimate is being created.

Effectiveness differs from one country to other, in regions inside a country also, towns inside cities. It can be impacted by the relative experience of the worker force, site administration including arranging, exchange union impacts, staffing conventions and legislative announcements on hours of work, wellbeing and so forth. With no particular experience of his own, an estimator may need to make various enquiries to relate the fundamental standards in his ownership, regardless of whether in-house standards or published standards, to the specific work condition, to represent the relative efficiencies of the current work. In any such enquiries, care ought to be taken with the goal that a full comprehension is gotten in regards to the data advertised.

Productivity and execution must not be befuddled. Productivity is a measure of the exertion required to deliver a predetermined outcome, while execution is the speed by which the outcome can be accomplished. Expanded productivity can enhance execution, yet execution can likewise be enhanced by expanded work levels, which thusly may have the impact of decreasing productivity and expanding costs.

It is worthwhile to notice that a productivity factor expressed as being more than 1 speaks to as diminishment in productivity while a factor under 1 is an increment. Productivity factor ought not be used with any locally gotten spending costs since these would mirror the local conditions. Examination of such spending costs with the organization standard may, therefore, give additional methods for approving a productivity factor.

Mostly, just direct worker hours are ascertained with indirect worker hours being adjusted as component of the gang rate connected to the quantity of direct worker hours. Direct worker hours are, some of the times translated as a contractual worker's own particular work when contrasted with subcontracted work. Other than this, circuitous worker hours might be comprehended as all work other than tradesmen and it is in this manner again important to characterize the data provided.

It is helpful to utilize a definition that has a progressing use for the duration of the life of the project, by arranging and cost control teams. All projects ought to be predictable to give a dependable database.

The definitions accompanying below are proposed for direct and indirect worker hours:

Direct work: artworks or exchanges that are straightforwardly required with the construction, creation, development, machining, and arrangement, and gathering, erection, completing and testing of the work. Direct work should incorporate all specialties and exchanges that specifically add to the development of a facility or that are of direct help and without whom the completion of work couldn't be accomplished.

Indirect work: work of a more broad nature that offers help to the immediate work and which is important for the deliberate culmination of the work; however which can't be allotted to discrete exercises furthermore, does not specifically add to the construction activities of a facility.

Indirect work may incorporate supervision, quality assurance and control, framework, cleaning, distribution center and transport facilities, crane drivers, material control, fire watches, security, dimensional control and construction camp operations.

ESCALATION, EXCHANGE RATES AND FINANCING CHARGES

On a noteworthy project, escalation and arrangement for variance in return rates can be an exceptionally huge segment of the estimate and it ought to be clarified in the estimate whether costs incorporate or prohibit recompenses for escalation.

Escalation

It is suggested to hold inside the estimate points of interest of the way in which escalation was computed, with the goal that any changes happening later in the project can be distinguished and detailed.

The estimate ought to incorporate points of interest of the base date of the estimate, the appropriate rates utilized and escalation remittance and the project program. The major contract parts ought to be plainly identifiable in a staged escalation program and foreseen use profiles delivered for the most noteworthy exercises. (Horngren, n.d.)

Sometimes, proprietors may foreordain the level of escalation to be accepted, on the premise that the escalation expected inside their computations with regards to the estimation of the product, will be freely loosely identical with escalation in the capital spending plan.

Currency fluctuations

The need to take into consideration currency changes relies upon the idea of the plant, the area of the plant, the length of construction and the nature of the proprietor's business. A few impacts of currency variances can be taken into consideration by the estimator, while others are out with the necessities of the capital cost estimate.

A case of the latter would be the utilization of a feedstock evaluated in a cash that is not quite the same as that of the product. This can influence the project in a way that is absolutely out with the control of the estimator and the project group.

The impacts of currency changes on the capital cost of the plant can be took into account in various diverse ways:

- Assurance can be acquired by purchasing forward currency, in which case the expected premium charge of such a buy ought to be incorporated into the estimate.
- The proprietor might accept the danger of currency changes, in which case the project will match with the proprietor a 'project contingency stipend conversion rate to be utilized for

the duration of the life of the project, with the proprietor going out on a limb of currency fluctuating from this set base.

- Proprietors that are multinational organizations with significant outside currency possessions may not be worried about currency changes since they may choose to pay remote cash accounts out of their outside cash property.

Financing charges

Funding for the financing of a noteworthy project all through the long times of design, development and construction can be equally vital. The estimator will make his income forecast by adequately evaluating the project plan by spreading the use against the periods over the time intervals over which that consumption is being brought about in accordance with a appropriate consumption use profile.

CONTINGENCY ALLOWANCE

A contingency allowance is a fund given inside an estimate to unexpected things, which experience demonstrates are probably required to cover design improvement, construction change and assessing alterations inside the project scope. Known prerequisites ought to be incorporated into the base estimate and not as a segment of the contingency allowance.

The contingency allowance excludes for drive majeure, or for the likelihood of particular scope alteration. For instance, the likelihood of needing piled foundations ought to be incorporated inside the body of the estimate, while the cost of unreasonably long piles ought to be suited by a particular stipend inside the contingency allowance. The likelihood of the necessity for extra piles in case of the plant capacity limit being expanded ought to be barred. (Horngren, n.d.)

Likewise, since it is probably the case that overall inflation can make costs rise, this ought to be accommodated inside the estimate under a particular heading; a stipend for startlingly high increments would be suited inside the contingency allowance.

The contingency allowance is not added to the base estimate to reflect the most elevated conceivable cost but instead the most plausible cost. It will along these lines not identify the extraordinary range of the exactness articulation.

The measure of contingency to be incorporated into an estimate, in spite of the fact that it is obviously subject to the data accessible, is also practically influenced by the information of the sort of plant being mulled over. An order of magnitude estimate, which depends on the investigation of an as of late built comparative plant, may have an exactness like a financial plan or conclusive estimate, and care ought to be taken to guarantee that any estimate that depends on real recorded cost is not punished by the expansion of an superfluously high contingency allowance.

As expressed already, the contingency allowance can be an exceptionally major saccount inside an estimate and accordingly should be taken care of in a way suitable to a noteworthy cost component. On a noteworthy project and where a person's experience permits, it might be important to present a likelihood investigation, or full hazard examination, of the different most extreme and least possibility figures set up, to give a practical aggregate remittance. (Salvendy, n.d.)

It is to be foreseen that evaluating precision will increment in line with changes in involvement, evaluating systems and assessing information. It ought to be recollected that upgrades that diminish assessing blunders or exclusions or all the more precisely anticipate cost or amount, ought to likewise enable a lessened contingency to be incorporated. Inability to modify the possibility remittance may bring about a high estimate and risk a project.

RISK ANALYSIS AND EVALUATION

A contingency allowance is a fund given inside an estimate for unexpected things, yet not for things, for example, force majeure or other extraordinary happenings that are absolutely outwith the control of any administration. On a noteworthy project it might be important to embrace a full hazard examination and from that point deal with the hazard through all the phases of the project.

ESTIMATE CONTENT

An estimate arranged for the design and construction of a mega project will contain many segments, more often called as accounts, which might be stated under headings. Over a process engineering project these would be normally as takes after:

		%	
1.	Design, procurement, project services	10.0	
	of which		
	• engineering design		6.0
	• procurement		1.5
	• project services		2.5
2.	Construction labour/supervision	30.0	
	of which		
	• civil		8.0
	• piping		14.0
	• electrical/instrumentation		6.0
	• paint/insulation		1.0
	• other		1.0

		%	
3.	Equipment	20.0	
	of which		
	vessels/exchangers		10.0
	rotating equipment		6.0
	• other		4.0
4.	Materials	20.0	
	of which		
	• civil		4.0
	• piping		8.0
	• electrical/instrumentation		4.0
	paint/insulation		1.5
	• other		2.5
5.	Site/construction overheads	5	
6.	Escalation	5	
7.	Contingencies	10	
	Total	100	

INTERFACES

While delivering an estimate the estimator will continually look to address the data given to guarantee that he has a comprehension of the full extent of the work and to fulfill himself that the data given by others additionally mirrors the full extension as he comprehends it. Cases of questions raised at this interface will incorporate the accompanying.

Process

Survey equipment list with the process specialists to guarantee that equipment not identified with the genuine process is incorporated, i.e. cranes inside a compressor building, and so on.

Procurement

Build up with the acquirement division that enquiries at check costs incorporate every expected component of cost, i.e.:

- Conveyance and documentation costs;

- Construction and initiating funds,
- Full painting specs
- Providers' site engineers
- Uncommon tests

Piping

Set up certainty level of material estimates delivered by piping engineers and accept any change important. Check with the piping estimating engineers a consideration for cutting and waste, which is a recompense for materials that must be obtained in overabundance of the introduced amounts, including remittances for short lengths of pipe not utilized, harmed, lost, and so on.

Construction

Construction probably conveys the most astounding level of hazard to an estimator what's more, care and perseverance must be utilized while evaluating construction costs.

The views of the construction office ought to be looked for the accompanying:

- Choice of proper construction standards
- productivity in connection to site area, level of pre-gathering, foundation area, development time, e.g. dominatingly winter or summer, working rises, nearness to other real development projects, working inside the bounds of live plants, necessity for working licenses, and so on.; work accessibility; site working permit including typical working week, prerequisite for working and productivity installments, sweep stipend reflecting neighborhood work permit, and so forth.

Monitoring change

When the conceptual design is finished and working at the spending estimate starts, it will be important to guarantee that progressions or potential changes to the estimate are recognized. Obviously with data at a preparatory stage and with various presumptions and approximations being made, it is at times hard to stay aware of the conceivable changes that can happen from any given or accepted base.

It is along these lines important to recognize the key reports and documents on which all different parts of design depend and to build up a predefined modification as the base and to watch change against that base. (Salvendy, n.d.)

There are five documented records that are basic to an engineering design, control of which will identify all significant changes:

- The premise of design or design detail
- The procedure/utility stream sheets and heat and mass balancing computations;
- The piping and instrument graphs and diagrams
- The equipment layout /plot designs
- The electrical single line diagram

Any change distinguished to these archives ought to be explored by the estimator and the cost implications evaluated by utilizing his own particular information or, on the other hand that of the design engineers, following which the fundamental documents ought to be raised or modifications made to the estimate.

CONTROL BY ESTIMATE

Expenses require control consistently yet control turns out to be logically more essential as the rate of consumption increases. Consumption will begin to quicken once the calculated design is finished and taken a toll data must be made accessible to set targets, record advance and consumption, foresee restorative activity and gauge last cost.

Cost that is being dynamically used over a noteworthy period can't be controlled against a solitary high worth section. In spite of the fact that the complete estimate may contain the required detail it won't be accessible at an early stage and if, for instance, the spending estimate contains a solitary figure for the mass purchasing of carbon steel pipework, this won't permit compelling control of the individual purchase requests to start.

The detail contained in a spending estimate may along these lines require growing or refining before it can be utilized as a compelling working device. In the event that data is not accessible to create an exact breakdown the estimator must make his best evaluation since even a self-assertive figure gives a reason for correlation and control.

CONTINUED USE OF ESTIMATE

An estimate ought not be set aside once the aggregate has been affirmed. The estimate ought to rather be a piece of a dynamic continuous premise of control of all genuine use or expected use with a specific end goal to best suspect what's more, control the last expenses. This control is to be practiced by taken a toll testing furthermore, cost control procedures.

CODING AND CONTROL

Keeping in mind the end goal to set up an arrangement of control amid the life of a project, implies of examination between projects, a formalized way to deal with the production of chronicled information, and a decent workable coding framework ought to be set up which perceives the different evaluating and cost control necessities.

Codes of records on project controls are also essential, yet their pertinence to evaluating is that they formalize reports into a structure that will encourage future estimates. Unfortunately, every client body has a tendency to have its own code of records and unless contractual workers run parallel systems, information caught on one project are not as effectively converted into another project, and PC use is in this way limited and less effective.

A decent coding framework will give the three or four best levels of detail furthermore, take into account the further detail that would be required by, for case, an enquiry in light of unit rates. It ought to likewise mirror a vocation execution approach since this gives intends to progress checking and control.

ESTIMATE PRESENTATION

The detail given in an estimate will rely upon the necessities of the beneficiary, yet regardless of whether the detail is to be presented, the estimate ought to be completely archived to give a premise to estimate corrections and reallocations and to give a premise to critical examination. Estimates delivered in a predictable way will aid the foundation of a reason for near examination and in-house assessing. (Salvendy, n.d.) It is, along these lines, desirable to utilize institutionalized forms, each dated, and containing obviously and compactly, all computations, suspicions and estimate premise. Obviously, the utilization of a PC in putting away and dissecting such information is of huge advantage.

CHAPTER 4

Planning and Progress Control

INTRODUCTION

One of the principal components of project control is the planning action.

Control requires a comprehension of the present status, a definitive objective and the activities important to accomplish that objective. This is as valid with regards to the general project as it is in the everyday exercises of each individual dealing with a project.

Planning itself includes all exercises required in setting up the procedure for a project; it impacts the technique for execution, the resourcing, the acquirement approach and the contracting system. The planning capacity proceeds all through a project, observing advancement, determining future patterns, measuring results, and announcing status.

THE IMPORTANCE OF PLANNING

The significance of project planning can best be shown through past project understanding. Many projects flop because of misguided judgments, underestimations of time and so forth in the underlying planning stages and due to the inability of administration to guarantee that the plan is clung to. It ought to be comprehended that planning engineers design the project and project administrators / managers deal with the project. It is accordingly the obligation of the planning specialist to set up the project exercises and occasional events, their coherent connections and interrelationships and the arrangement in which they are to be proficient, which is the course outline by which the project supervisor will coordinate the project. The inability of the planning engineer to create the redress outline or the project administrator to take after its necessities will have desperate outcomes. (Salvendy, n.d.)

The amendment of issues might be nearly easy if distinguished at an early stage; yet as a project advances the cost of redressing mistake or, then again integrating change increments significantly. The fulfillment date is too progressively at risk and the noteworthy absence of time in the later phases of a project can bring about limitations, or adjustments to the favored methodologies, uneconomical working and extra cost.

If work can't be foreseen it can't be arranged, if work can't be arranged it can't be controlled and will be done in a specially appointed way with costs being observed and not controlled. This remark applies similarly when "time" is replaced with 'cost'.

The development of an appropriately considered plan is fundamental to the business accomplishment of a project and should be delivered by the individuals who have understanding of a comparative kind of project, and who value the full scope of conditions and connections between the different exercises which must be performed with a specific end goal to finish the project. (Alistair Barros, n.d.)

ORGANIZATION

To carry out a perplexing project effectively, it will be important to set up a devoted, mindful, planning group giving an exhaustive planning, checking and control benefit that assembles and disseminates data by the receipt and issue of suitable reports through the entirety of project, from the proprietor to the project group and through the design and construction contractual workers to the different sub-contractual workers and providers. The level of straightforwardness and control that can be practiced at any level will rely upon the authoritative connections.

It will be of vital significance to the project that the planning engineer guarantees that a state of 'no surprise' exists for the duration of the life of the project.

While the planning action grasps the time traverse from the occasion a project is conceived through to its consummation, the begin purpose of a specific planning activity will rely upon whether the organizer is working for a proprietor, contractual worker or sub-contractual worker. An organizer taking a shot at benefit of a proprietor will begin with a clear sheet of paper and will be controlled as it were by the level of desperation communicated by his administration. Organizers working for contractual workers or sub-contractual workers, then again, will work inside expanding imperatives the lower they are on the contracting pyramid. (Alistair Barros, n.d.)

The control of a perplexing project may require the proprietor's project group to demand that every single contractual worker taking part in the project should utilize planning frameworks that are perfect with the larger amount prerequisite, both as system and framework programming. A similar group may expect data to be given on advance accomplished, foreseen finishing dates for the work or parts thereof, and maybe keeping an eye on levels to guarantee the smooth movement of work starting with one contractual worker then onto the next. Also the group may require planning and advance frameworks to be good with work bundles and other work breakdown structures to give mix of reporting. Such necessities can, obviously, profoundly affect contract procedures and connections.

LEVELS OF PLANNING

The level of detail required by the different workforce or associations doing work at a project relies upon their position inside the project. What might be considered as simple detail to the proprietor might be the full extent of work to a sub-contractual worker.

Similarly as with different regimes of services associated with a project, planning frameworks and reporting necessities are planned in a chain of hierarchy, each level of the hierarchy order containing point of interest suitable to the work being controlled and level of administration and services included. The level of detail important to control the day-today exercises is not required by a project supervisor / administrator evaluating the general status of the project, while corporate administration may just be intrigued in a high level 'official synopsis'.

Different levels of planning and reporting are along these lines important to mirror the prerequisites of the clients' data, with each level being equipped for cross examination at a more prominent level of detail through the progressive hierarchy structure.

The planning capacity initiates by the setting up of an arrangement for the general project as a Gantt or bar graph. The project program is then arranged which sets the arrangement in a periodic scale and assesses key actions dates and turning points. Point by point / detailed control programs are at that point created in bar diagram plans got from logic network systems logic network systems recognize action time spans, the coherent connections amongst actions and exercises and take record of the valued parts of the required and accessible assets.

At last, singular task level calendar schedules are readied, in view of the control program, for the everyday control of the physical work. Responsibility for design is not kept to the planning designer and it can't be underlined too firmly that the counsel and cooperation of the parties that will be working the plans, e.g. engineers, purchasers, construction supervisors / administrators / managers, and so on ought to be requested. All individuals from the project group ought to partake in the definition of the arrangement and planning from its beginning and complete 'buy in' to the calendar in its last frame. (Alistair Barros, n.d.)

Physical work will be communicated regarding physical advancement accomplished. Advancement made with design and comparative exercises will as often as possible be communicated as deliverables. A deliverable is the physical consequence of the fruition of a discrete bundle of work and typically relates, however is not restricted, to drawings, particulars, systems, documents, reports and different archives.

Illustrated beneath are the four fundamental levels of planning, there might be extra levels relying upon the idea of the project. Levels 1 and 2 will be created by the proprietor's project group and levels 3 and 4 by the contractual worker, who may likewise have his own particular adaptation of levels 1 and 2.

Level 1 bar chart

The level 1 bar graph is the most essential graphical show of work done drawn against the time. It will demonstrate the general program for culmination of the project as a period based bar graph (now and then called a Gantt chart) of maybe 30 exercises and in single sheet organized format.

Such an outline, a streamlined form of which is appeared in Figure, gives a general perspective of a project's necessities with finished work maybe spoke to as a solid line and extraordinary work as a broken line contrasted with a 'period now' line to demonstrate exercises ahead or behind program. A bar graph can't demonstrate the unpredictable interrelationships and conditions between the numerous exercises in a project and is not a report against which point by point detailed control can be practiced since it doesn't contain the data important to screen advance in controllable bits. (The Joint Development Board sponsored by the Association of Cost Engineers and the Royal Institute of Chartered Surveyors, n.d.) The level 1 program will ordinarily be utilized as a rundown by senior administration and may likewise contain synopsis advance data and key dates as appeared in Figure.

Key dates, once in a while alluded to as turning points, will normally have a single deliverable for their accomplishment, for instance:

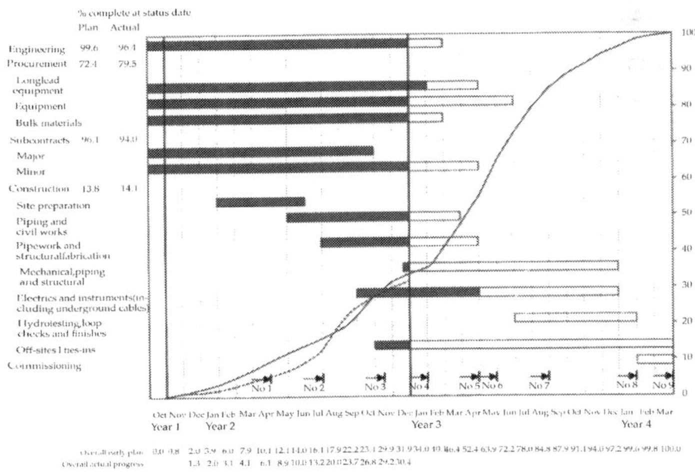

Figure 4 1: Level 1 bar chart with key dates (milestones) and progress curve (The Joint Development Board sponsored by the Association of Cost Engineers and the Royal Institute of Chartered Surveyors, n.d.)

- Arrangement of procurement orders for specific things of equipment;
- Activation of a key contractual worker;
- Culmination of detail design;
- Conveyance to site of certain predefined things of equipment;
- Culmination of substantial lifts;
- Power up;
- Mechanical consummation.

The key dates will come about because of the more point by point contemplations in the level 2 program.

Level 2 program

The level 2 program is the most astounding working level of the planning structure it is the level at which the connections between the different exercises are distinguished and will be the project administrator's working apparatus. Since it will distinguish all key exercises and key dates it will ordinarily be the project administrator's 'master schedule' and can be utilized to track foreordained project points of reference and the related endorsements required from the proprietor.

This program will be as a period based connected bar graph of maybe 100 key exercises and phases of finish like that appeared in Figure below. All dates appeared on this program having been determined from the dissected level 3 plan. The level 2 program will demonstrate the general project status and will be refreshed and reissued at customary interims appearing, in respect to a vertical dateline, the advance planned and accomplished for every exercise.

Level 3 network

The level 3 program is typically seen as a bar graph determined from an intelligently connected network. A network distinguishes all exercises important to the finishing of a project, the time required for every movement and the consistent reliance of exercises on each other. For each network there will be a chain of intelligently connected exercises from the beginning of the network through to the completion of the project that characterizes a term through the project in general which cannot be lessened. This is known as the critical path in light of the fact that any invade on any of the exercises on this critical path will cause an overwhelm of the project.

A first go at the level 3 program, as a evaluation network, will most likely be done by the proprietor's project group, yet the "complete" level 3 program is generally delivered by the contractual worker as his point by point control program. This program covers the whole extent of the contractual worker's work as a sensibly connected network, recognizing the key dates, expectations, interfaces, material acquisition and subcontracting exercises.

In delivering the "conclusive" level 3 program, the contractual worker's planning engineer and designer will assess every single "gentler" connection, for example, asset

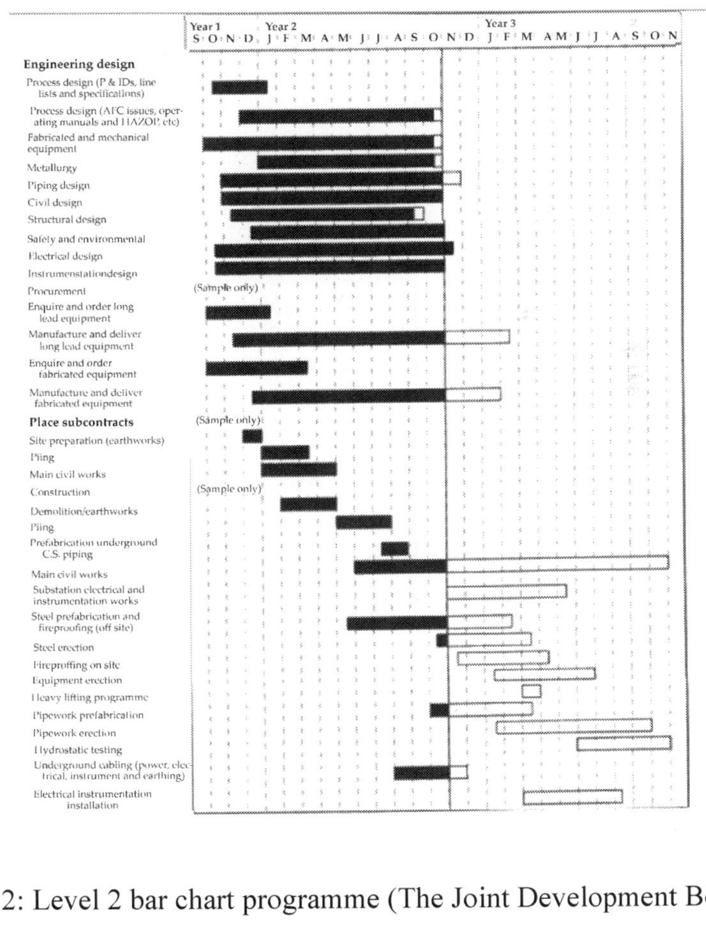

Figure 4 2: Level 2 bar chart programme (The Joint Development Board sponsored by the Association of Cost Engineers and the Royal Institute of Chartered Surveyors, n.d.) and labor restrictions, which are under administration control, when contrasted with sensible connections which are definitely not. In this manner, for instance, the stripping of shutters is completely reliant on the curing of cement, yet it might likewise be subject to the accessibility of craftsmen. The main connection can't be changed, though the second can.

The program will take into consideration the necessities of any noteworthy key dates in the 'project date-book' and when the examination is performed, it computes the beginning and fulfillment dates, assessing non-working days and 'aggregate float' for every movement. Float is the measure of over run on an exercise that can be obliged without it influencing the project consummation date. Movement and activity spans are normally appeared in days. At the point when the itemized design has been produced and the investigation run and accepted, exercises are stacked with worker hours and preparatory labor histograms delivered. These are looked into with the lead engineers of various disciplines involved, construction supervisors, and so on and exercises are moved inside their drift constraints to get the ideal "smoothed" histogram, i.e. a work profile without over the top pinnacles and troughs. Now the program is solidified, frequently alluded to as being 'base lined' and the level 2 program also, level 1 bar graph are then created.

The level 3 program is the premise against which those endeavor the work report advancement and likely finish dates for every action, utilizing summated data acquired from level 4. The level 3 program will generally be refreshed fortnightly amid design expanding to week by week amid the development stage. The refresh will be founded on data got from engineers, construction administrators, and so forth showing accomplished advancement also, current figure against the arrangement for every action. (Tonchia, 2008)

The contractual worker ought to be enabled the opportunity to design and deal with his claim work and unless the proprietor is completing a portion of the work himself, level 3 will be the most reduced level at which the proprietor's project group will work on an everyday premise. The proprietor will require such detail as will permit him to deal with the different interfaces, yet should oppose the allurement to design at too low a level and in an excess of detail. (The Joint Development Board sponsored by the Association of Cost Engineers and the Royal Institute of Chartered Surveyors, n.d.)

Level 4 schedules

The level 4 schedule changes over the theoretical aspects of the network into the fundamental level contemplations for executing the work. Level 4 data for the most part involves registers containing all the significant points of interest of requirements or site work relating to the work components inside each level 3 action and are the reason for the ensuing everyday planning, checking and control of work scope and physical advance. The level 4 schedules are created by or in the interest of the contractual worker or whoever is executing the work.

Each level 4 design deliverable or component of site work, recognized by the design orders or construction administrators, will be relegated a code number which will interface the records or work to exercises at level 3. It is normal for the exercises at level 4 to be coded and composed to be equipped for being summated in various distinctive ways. For instance

JOBCARD NO	DISCIPLINE		PAGE NO	
CONTRACT NO REVISION NO PLANNED START COST CODE	ISSUE DATE REV DATE PLANNED FINISH		DURATION	HOURS
JOB DESCRIPTION e.g. fabricate and weld bon A jacket leg A1	RESOURCE TYPE	RESOURCE QUANTITY	RESOURCE DURATION	
DRAWINGS REQUIRED				
SERVICES REQUIRED e.g. SCAFOLDING RIGGING	TOOLS, NDT	WELDING DOC REF NOS		
PERMITS REQUIRED HOT WORK	VESSEL ENTRY		ELECTRICAL	
	CONTRACTORS SIGNATURES			
NSPECTOR DATE			SUPERVISOR DATE	
	SITE MANAGEMENT SIGNATURES			
NSPECTOR DATE			SUPERVISOR DATE	

Figure 4 3: A Job card (The Joint Development Board sponsored by the Association of Cost Engineers and the Royal Institute of Chartered Surveyors, n.d.)

albeit at first the reporting and planning of level 4 exercises might be on area premise, at a later stage it might be important to deliver data arranged on a framework premise in order to give a nitty gritty report on an account of the exercises also, worker hours required to finish each of the procedure frameworks without regard to area.

Shutdown planning

At the point when a process plant is shut down for repair, support or updating, construction exercises are arranged and controlled vivaciously because of the degree of the work to be attempted over a short contract period, which regularly requires 24- hour shift work.

In these circumstances work cards like that appeared in Figure might be utilized. The work card will contain, for instance, the level of detail important for a field supervisor to arrange his workforce, since it will designate the timing and assets assigned for the fulfillment of every movement. It will likewise convey points of interest of all drawings relevant to the work.

NETWORK TECHNIQUES

The network is at the core of the planning capacity in that it contains the rationale on which the different planning plans are based.

A network comprises of a succession of occasions intelligently connected together and contains data in regards to the type of occasions, the order in which occasions can be accomplished and the term of any activity important to their accomplishment. The span of a path through a network is accordingly reliant on the term of the constituent exercises, while advancement is plotted against occasions.

There are two traditions of speaking of networks. The first tradition advocates exercises as arrows, the occasion being the hub called node at which the arrows meet. The second is named as precedence network which shows the exercises as the nodes of the graph which are associated into a chain of rationale by arrows known as constraints. The first is now and then alluded to as 'activity on arrow', the second 'activity on node' which is the present hone.

CRITICAL PATH AND FLOAT

Keeping in mind the end goal to ascertain the soonest / early time of an occasion, it is important to consider the time frames of exercises that go before it. This is known as a forward pass.

Keeping in mind the end goal to ascertain the latest time that an occasion can happen while not influencing the general finishing date, it is important to consider the spans of exercises between the occasion and the finish of the project. This is called a backward pass.

The forward and back passes through the network create for each occasion a scope of times at which the occasion can happen without risking the project finishing. The breaking points of this range are the most early conceivable occasion time, created by the forward pass and most late conceivable occasion time, created by the backward pass. Where the most early and most late dates agree these exercises are known as basic exercises and the basic path is that which interfaces these basic exercises.

At the point when the forward and backward passes have been finished an end date for the project winds up noticeably clear. By settling this end date any future investigation broadening basic exercises will be demonstrated as negative float,

which features the culpable exercises to the planning engineer. To hold the original contract finish dates it will be important to audit the (presently) supercritical exercises and maybe increment manning levels to lessen the time frame of the exercise, or deal with the issue by featuring the trouble and giving it customary particular consideration.

Exercises might be on the basic path at a specific time in the project, but may from that point stop to be basic, because of a specific incident, or absence of happening; also, exercises that were not on the basic path may wind up noticeably basic.

Occasions far from the critical path have a scope of accessible circumstances ascertained by the forward and back pass, the contrast between the most early and most late circumstances for an occasion is known as float. The critical path will be the path with the minimum float, regularly zero.

For a meaning of the terms utilized as a part of project networks reference ought to be made to British Standard (BS 4335) entitled

Glossary of Terms utilized as a part of Project Network Techniques.

PROGRESS MEASUREMENT

All together for the project group to estimate the last consummation date and cost, a basis / premise for examination should be set up between the expected and genuine advancement and expected cost and cost to date.

The level 4 plan shapes the premise of the contractual worker's advancement control framework. It is at this level the contractual worker will record advancement, distinguishing what has been accomplished and when it was accomplished. The genuine hours utilized as a part of its accomplishment will be recorded against time sheets which are typically what might as well be called level 3 data.

This data will be utilized to figure worker hours required to finish the work considering advancement and productivity accomplished to date, by area, discipline or framework. (The Joint Development Board sponsored by the Association of Cost Engineers and the Royal Institute of Chartered Surveyors, n.d.)

Advancement is not measured by looking at time used against total time permitted, nor is it quantified by contrasting worker hours used and total worker hours foreseen, or by looking at sum paid and total sum of the contract. Advancement is an examination between work expected and an estimation of real physical work accomplished. Advancement may be measured against "coordinate" or "progressable" exercises, i.e. exercises that specifically add to the production of the unit. Administration and management support exchanges, for instance, not measured as advancement.

Calculation is taken out intermittently at periodic interval dates, the strategy for physical calculation of advancement fluctuating between disciplines. Obviously the strategy for measurement and calculation of design, acquisition and comparative exercises will vary from that utilized as a part of association with construction works where physical advancement is specifically quantifiable.

All together that advancement on design and comparable exercises does not remain a matter of assessment it is important to agree, before initiation, the arranges through which expectations, e.g. format drawings, general plan drawings, enquiry reports, and so on will continue keeping in mind the end goal to build up a system against which advancement can be assessed. This is more often alluded to as benchmarking. The accepted stages ought to contain a depiction of the "quality" of the deliverable required, as advancement on a deliverable containing different "holds" or spaces is faulty.

Original design exertion and advancement depend on the designer's estimate with regards to the sorts and amounts of documents and reports to be created and worker hours built up for the diverse stages through which the record must be prepared. Advancement is evaluated by the specialists at the most reduced level of detail, e.g. drawing, determination, assessment and study, and so on. The physical status of the report document is checked by utilizing the benchmarks set down for the stages, per record and the specialist's own judgment.

Procurement can be arranged and checked by setting an acquirement cycle which incorporates such exercises as enquiry issue, order issuance, and so on.also, designating worker hours in like manner. Time designated to each phase of the cycle assesses seller (provider) documentation necessities, readiness of design deliverables, etc. Procurement advancement is measured against the Procurement cycle per demand for the aggregate Procurement exertion.

Design projects will normally be construction driven, in that the necessities of the huge construction workforce will take need over the bother of the designers. The conveyance dates for acquired things will be connected to the construction necessities; equipment and materials will in this manner be 'figured out' to assess when it is required to suit site exercises. Certain equipment will be basic with respect to seller (provider) information as opposed to conveyance, in which case the enquiries will be customized early in the project in order to be in receipt of information to suit design prerequisites.

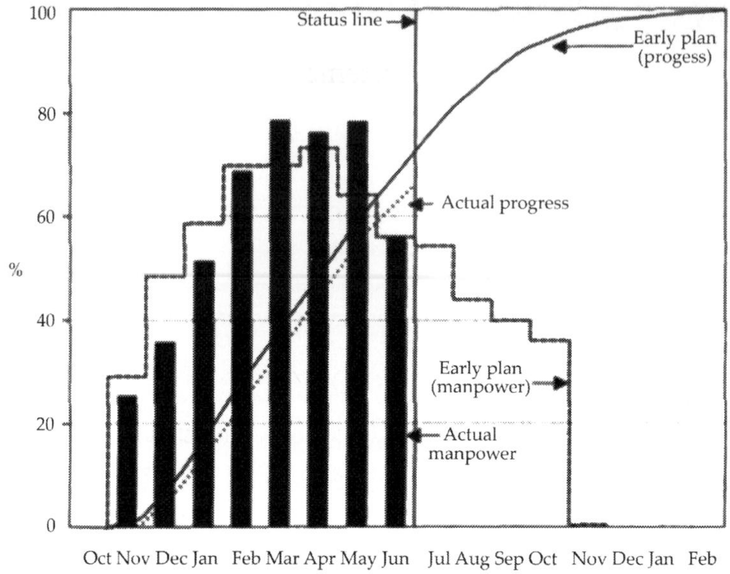

Figure 4 4: usual curve of level 2 progress (The Joint Development Board sponsored by the Association of Cost Engineers and the Royal Institute of Chartered Surveyors, n.d.)

Data got from the physical advancement estimation will be utilized in conjunction with the "S" curve and planning breakdown to indicate real physical advancement against scheduled and plannned advancement on a given date (see Figure).

Measurement made for the reasons for reporting advancement is most certainly not essentially reasonable for different purposes, i.e. accepting accounts, despite the fact that it is every now and again utilized as a reason for the understanding of break installments by the revealed advancement being connected to the present estimation / measurement of a component.

Histograms and progress 'S' curves

Histograms demonstrate the way in which work levels intermittently increment and lessen over the life of a project or part thereof. They fall into two classifications, progress-able and non-progress-able as illustrated below:

1. non-progress-able:

- Administration and management support services;

2. progress-able:

- Engineering design;
- Acquisition;
- Supervision;
- Staff;
- Work in terms of professional skills

There may likewise be a rundown histogram showing general work.

The contractual worker is generally required to submit labor histograms for the cooperate with curves ('S' curves) demonstrating graphically the expected advancement versus real advancement all through the agreement time frame.

The "S" curve will demonstrate the planned and original percentage consummation for each revealing period and will be produced from the level 4 program. The underlying planned curves as a rule stay unaltered for the length of the agreement unless there is a noteworthy change that would some way or another make correlations unimportant.

Histograms can be obtained by applying assets to each network exercise and totaling the outcomes per period with a specific end goal to acquire required work totals to week after week, month to month, or to whatever other period might be picked.

Mechanized network programs complete the total as well as additionally plot the subsequent histograms. Histograms may reveal erratic labor work levels that may demonstrate that the network rationale requires consideration.

CONTINUING CONTROL

Once the project is under way the planning specialist will frequently audit also, write about the status of the work and will refresh the planning database. The planning specialist will facilitate his exercises with the cost and contract engineer to guarantee that all project controls can be utilized in conjunction of each with the other. This is especially imperative concerning the integration of accepted changes into the extent of work. Data will be gathered on man-hour use and advancements accomplished for examination with that planned. The planning specialist will, together with the material controller, screen conveyances of material and equipment against planned and original prerequisites together with the issue of design data and the arrival of any "holds" on design drawings.

Progress reports

Every contractual worker effectively included in the project will create advancement reports at normal stipulated periods, as per the prerequisites of the project group. The project group's planning engineer will then deliver a united advancement report for the entire project.

A regular month to month advancement report needed from a contractual worker will include the things mentioned below.

Summary

A summary of the present status of engineering design, procurement, contract arrangement, equipment and material conveyances and construction advancement. Particular reference ought to be made to any present issues, bottlenecks, conveyance slips, and so forth. the impact on program andlor costs, and a sign of the means being taken to beat them.

Achievement

An illustration and description of advancement and summary of the work force in terms of professional career or discipline for both the contractual worker and his sub-contractual workers.

Objectives

This includes a depiction of exercises to be completed in the following period.

Areas of concern

Features any issue areas that may make postpone to a key date or to the consummation of the work.

Pending or approved change orders

A synopsis of pending or endorsed change orders got from the cost and contract engineer.

Accidents and losses

An outline summary of mischances and misfortunes as created by the safety officer.

Planning matters

An outline summary of advancements demonstrating any deviations from the agreed program and their conceivable impact on general finish of the work.

Bar graph or other graphical representation to demonstrate planned and real advancement for:

- Engineering design;
- Procurement;
- Material conveyances to site;
- Construction

The graphical representation is as often a histogram based on original control worker hours and the first project program and demonstrates current labor necessities, generally and by discipline. The premise of the histogram will be refreshed occasionally in light of reexamined worker hours and agreed reconsidered program.

The cost engineer will utilize data associated with advancement accomplishment when determining last cost and creating different insights; it is imperative that cost and planning controls are kept up on a practically comparable premise.

Intermediate reports

On significantly large projects or amid basic phases of a project it might be essential for a contractual worker to be required to create week after week or even day by day reports.

Such reports might be limited to the accompanying subjects:

- Details of the real contrasted with the planned physical

advancement of the work accomplished amid that period;

- Keeping an eye on levels by significant exercises;
- Materials and equipment status, demonstrating any areas of prime concern;
- Any issues which could imperil opportune culmination of the work;
- Exercises to be completed amid the accompanying time frame;
- Climate report;
- All downtime;
- Mischances and accidents

CHAPTER 5

Cost Material and Document Control

COST CONTROL

Objectives

The Association of Cost Engineers characterizes cost control as 'The procedure of

- Controlling all cost factors associated with a task so that production for a characterized obligation are accomplished at a reasonable cost inside the measure of cash appropriated'. Keeping in mind the end goal to accomplish this, cost control exercises on any project ought to envelop the below objectives:
- Recognize potential cost inconvenience spots in time for restorative or cost decreasing move to be made, i.e. identify potential spending overwhelms before they turn into a reality;
- Guarantee that the individuals who are spending the project reserves know about the financial plan for their zone of obligation and how their consumption execution is related to

the financial plan;

- Make a cost-cognizant environment so all in the project group are cost cognizant and mindful of how their exercises affect the project cost;
- Limit project costs by taking a gander at all exercises from a cost lessening perspective.

To achieve these targets it is important to:

- Have a reasonable budgetary measuring stick (the control estimate);
- Separate the control estimate into controllable bundles of work
- Using code of records and additionally work breakdown structures as required;
- Produce exact and convenient cost estimates for these bundles of work;
- Contrast these estimates with the measuring stick with distinguish potential issues;
- Make positive move to limit these potential issues.

The foundation of a control is the making of an agreed and accepted characterized base, the checking of results against that base, the recognizable proof of antagonistic patterns, the proposing and executing of remedial measures and the observing of such measures to guarantee that they have the coveted impact.

Obviously the distinguishing proof of unfriendly patterns must be in adequate time to enable remedial moves to be made, and along these lines cost control is a progressing ceaseless action as opposed to an infrequent occasion.

It ought to be noticed that cost control can't be seen in confinement. Cost and time are interrelated where an overwhelm on time frequently brings about a increase in costs. It is consequently essential to consider advancement, estimation of work done, productivity factors, and so on when determining a cost prediction / forecast.

Cost control does not really imply that a task is done in the least expensive conceivable way, yet it ought to guarantee that the project will be finished at a financial cost inside the accepted estimate. It ought to likewise guarantee that there is no overdesign or uneconomic design.

Cost control terminology

It is important to see a portion of the terms as often as possible utilized as a part by cost control and by cost engineers. Perplexity regularly exists with regards to the contrast between duty, gathering and use, point by point /detailed records of which are kept up by cost engineers for correlation with a control estimate. These terms do have a tendency to fluctuate in minor routes over the business, yet a regular definition is sketched out beneath.

Commitments

Responsibilities against a financial plan inside the control estimate will typically include the accompanying:

- Worth of orders incurred
- Faxes or letters of intent provided
- Estimated worth of orders with no settled worth
- Remittance for escalation (which will incorporate cost of cargo, carriage, obligation and expenses)

Accrual

The Accrual is the variation between the invoiced worth and the real worth of work performed, be it design engineering or construction and establishment work, at a given point in time. While to some degree an estimation, the invoiced worth together with an accrual is typically a sensible evaluation of the worth of work done. It is accordingly the real device utilized for determining as duties are recorded ahead of time of work really being done and consumption / expenditure lingers behind.

Expenditure

Consumption / expenditure is a record of 'money out the account'. Because of the time slack between the work being done and the receipt for that work really being paid it is of little use to the cost engineer as an anticipating instrument. Be that as it may, it is of evident significance to the proprietor's bookkeepers who require to know the amount of project sums which will be required on a month to month premise to meet receipt installments. As a feature of his duties the cost engineer is frequently required to supply income figures all the time amid the life of the project.

The control estimate

The foundation for keeping costs in control is the control estimate. The control estimate is created by taking the best estimate accessible at project initiation, much of the time the conclusive / definitive estimate, and arranging it into related classes which mirror the route in which the project is to be executed. The entire cost control rationality depends on the control estimate in that all predicted costs determined on the project are contrasted and the proper estimate recompenses contained in this record document.

With a specific end goal to guarantee that these contrasts and comparisons are on a strong basis the control estimate ought to be separated in a way which will be good with the way later data will be accessible. The control estimate ought to be appropriately fit the bill to show to users and clients the level of precision which is employed in its preperation and access to the full record of data utilized and presumptions made in its planning must be granted.

Before the control estimate being set up it is obviously still fundamental to control costs bearing against the project. These costs will fundamentally comprise of plausibility / feasibility and front end engineering exercises and should be taken into account when assembling the control estimate. Once the control estimate is accepted, all issues which may affect last cost will be under steady survey including changes, forecast of cost, time, assets and contingency allowances required to finish the project.

Changes to the control estimate

Maybe the most essential part of cost control in the engineering department or, on the other hand on the construction site is the control of changes. The incite recognizable proof of changes to the extent of work and the fast transformation of these changes into time and money related terms, where appropriate, is of indispensable significance in the control of a project. Without these, control and cost gauging frameworks can't mirror the real extent / scope of work. As extent of work changes are recognized, estimated and approved they should be added to the project control estimate guaranteeing that the present control estimate dependably mirrors the agreed current extent of the project. The first control estimate was finished at a particular point in time and mirrored the project scope as it remained by then. By the expansion of scope / extent changes the estimate is stayed up with the latest, continually mirroring the present project scope.

Alterations to design are regularly unavoidable because of the idea of plant design. It is, however, fundamental both to an effective design operation and to successful cost control that design changes are controlled and decreased amid the itemized design stage, opposed amid the production design stage and acknowledged just in extraordinary conditions amid the development stage.

Changes should in this manner be recognized early and conveyed to the consideration of the project administrator / manager and proprietor quickly all together not to disturb the design exertion.

All design archives exude from a couple of key documents and reports which framed the premise of the estimate for the project. It is unrealistic to fundamentally change the design without changing all or one of these key reports. Along these lines, if changes to key archives are controlled, changes to the design can be controlled.

It is fundamental while considering changes to compare accepted changes and non-accepted changes when detailing. An accepted change is one which is acknowledged by both proprietor and contractual worker as being important to the project and for the contractual worker to have a privilege under the agreement. Such changes would be added to the first control estimate and consolidated into the present control estimate and project plan. Along these lines, expecting productivity is as of now loo%, predicted finalized cost would be present expenditure divided by the advancement to date.

Non-accepted changes may incorporate costs for which the contractual worker has no appropriate to repayment under the agreement and contract based changes for which agreement and acceptance is being awaited. The first control estimate ought not be altered for such changes and they ought not be integrated into the current control estimate or project plan. Cost and time has or will be consumed on the "change" and this must be considered while computing the predicted last cost.

The contrast is critical when reports / documents are seen by the proprietor also, the contractual worker. These non-accepted changes, while by and large being dismisses by the proprietor as legitimization for extra repayment, do feature factors which may have unfavorably influenced the contractual worker's execution. These elements in this way may well shape the premise of future claims to be held up by the contractual worker against the proprietor.

It is common to integrate just agreed changes into the present control estimate and project plan. To guarantee that cost and planning are continuously on a similar premise it is imperative that consideration between contractual worker and proprietor is such in order to give a fast reaction against change orders.

Cost segregation

The spirit of any cost control framework is to separate the project into bundles of work that can be controlled and overseen. A project can't be controlled by just monitoring the aggregate consumption on it. After all things considered, utilizing this strategy, the project group would just acknowledge they had an issue when the aggregate project sums ran out.

On the off chance that the cost engineer is to distinguish potential issue ranges and severe patterns it is vital for a project to be separated into a progression of spending plans which can be exclusively announced against. These financial plans ought to reflect both the cost components of the project, i.e. material and equipment buys, contractual worker costs, design costs, and so on., and the route in which the project is to be executed, along these lines making it conceivable to allot spending plan duty to the suitable individuals from the project group. (The Joint Development Board sponsored by the Association of Cost Engineers and the Royal Institute of Chartered Surveyors, n.d.)

These two contrasting prerequisites are fulfilled by the appropriation of two interrelated revealing structures generally alluded to as the work break- down structure (WBS) and the code of records / accounts. The cooperation of these structures permits the revealing, reporting and control of costs by discipline or by area or, on the other hand by discipline inside a specific area, e.g. aggregate cost of labor work, aggregate cost of all necessities in an area and aggregate cost of labor work in an area.

At the point when these two sorts of structure are utilized with a corresponding coding framework the outcome is an adaptable reporting and revealing matrix equipped for delivering considerable detail if required, while abridging at different levels and in different organizations to suit the prerequisites of the beneficiary.

Work breakdown structure

A project needs the strict control of time and cost and consequently a progressive hierarchical structure is required which is relatable to these two basic foundations of control. As beforehand expressed, control not just includes observing original costs additionally the comparison and contrast of such costs against a settled base and the auspicious recognizable proof of unfavorable patterns and prescribed remedial activities.

The chosen structure is usually alluded to as a work breakdown structure (WBS) and encompasses the aggregate project with each layer of the progressive hierarchy having progressively more noteworthy degrees of detail. It will move from proprietor, contractual worker and sub-contractual worker exercises and will connect characterized bundles of work, their estimated and original cost and their estimated and original timetable. Such bundles of work at whatever level in the WBS might be alluded to as work bundles, in some cases as Cost Time Resource - CTR bundles or Activity control packages - ACP or different titles; however they are generally comparable in idea.

On most projects the individual who most impacts design advancement and cost is the design engineer at the working side or his supervisor, who might be directing various engineers designing a specific area of the work. On a construction site the administrators may have the most noteworthy impact on advancement and cost.

The individual who could have the greatest impact on a bit of the work ought to likewise have obligation regarding that part of the monetary allowance as far as both time and cost. Most issues begin at the detail level and most arrangements can be found there and the control structure ought to hence not just achieve this level of detail however should deliver data, for example, advancement to date, cost to date, and so forth., which will help the mindful people in effectively meeting their duties. (The Joint Development Board sponsored by the Association of

Cost Engineers and the Royal Institute of Chartered Surveyors, n.d.)

A regular work bundle will contain the accompanying data which serves to characterize the financial plan against which control is to be worked out:

- Extent of the work bundle
- Name and title of the engineer responsible
- Planning network details and timetable information
- Asset worker hours and cost synopses
- Anticipated cost of materials to be used in bulk
- Anticipated cost of plant and equipment

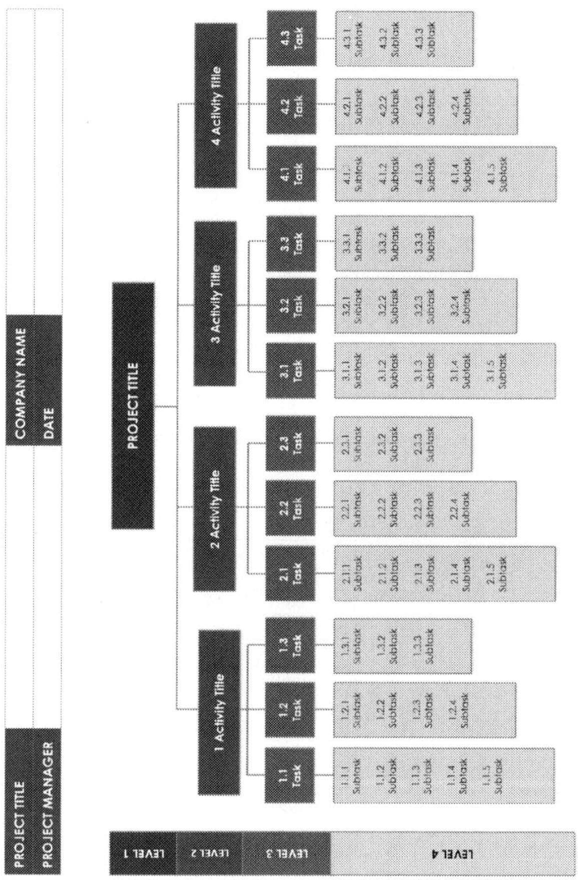

Figure 5 1: A Typical work break down structure

Work bundles ought not be so huge as to make the distinguishing of issues and problems unduly troublesome, nor should they be small to the point that trouble is experienced in directing a substantial number of bundles with their different interfaces. Likewise remembering that the most exact measure of advancement is the point at which an exercise / activity is 100% finished and as severe patterns should be distinguished early, work bundles ought not be of an extreme span.

Thought ought to be given to the level at which control is to be worked out. It might for example be viewed as important to screen advancement on singular drawings just to screen man-hour consumption against advancement on a bundle of drawings.

In choosing what level of control is most proper the accompanying criteria ought to be taken into consideration?

- Precision of base data at the most minimal level of detail;
- Measure of time required to keep up a profoundly nitty gritty framework
- Potential for blunder / error increments in extent to detail

It is not ordinarily suggested that worker hours consumed per document / report is recorded. Such records create over the top detail as well as broadly fluctuating outcomes because of the correlation of exact originals with an estimate at its minimum exact level.

For a Work breakdown structure to be efficient it must be organized in a way which will enable data to be 'moved up' to each progressive level of the chain of command; the best levels will accordingly be proprietor "claimed" and the lower levels contractual worker or sub-contractual worker 'claimed'. The proprietor should characterize his prerequisites to the contractual worker, subsequently permitting the accumulation of results at the interface. The contractual worker should guarantee a comparable course of action with his sub-contractual workers.

Code of accounts

The code of accounts is now and then alluded to as a cost breakdown structure, since it reflects cost classes, and is in some cases alluded to as an organizational breakdown structure -OBS, since it reflects to a degree an organization's authoritative structure.

Figure 5.2 below is dependent on the standard code of accounts which are distributed and published by the Association of Cost Engineers which characterizes a code of accounts as 'An intelligent marking framework using a progression of numbers and letters to allow the costs of like or related things to be gathered together'. This naming framework gives the premise to give an account of use, duties and expectations of costs to the finish of the project. The essential reasons for implementing a code of accounts framework can be illustrated as below:

To total information legitimately;

- To implement a well-established way to deal with information accumulation;
- To aid the control of costs;
- To give cost information to future projects.

Coding frameworks change over the construction sector however regularly comprise of a progressive framework making revealing and reporting at different levels of detail conceivable. Along these lines they guarantee that point by point / detailed data can be extricated from the

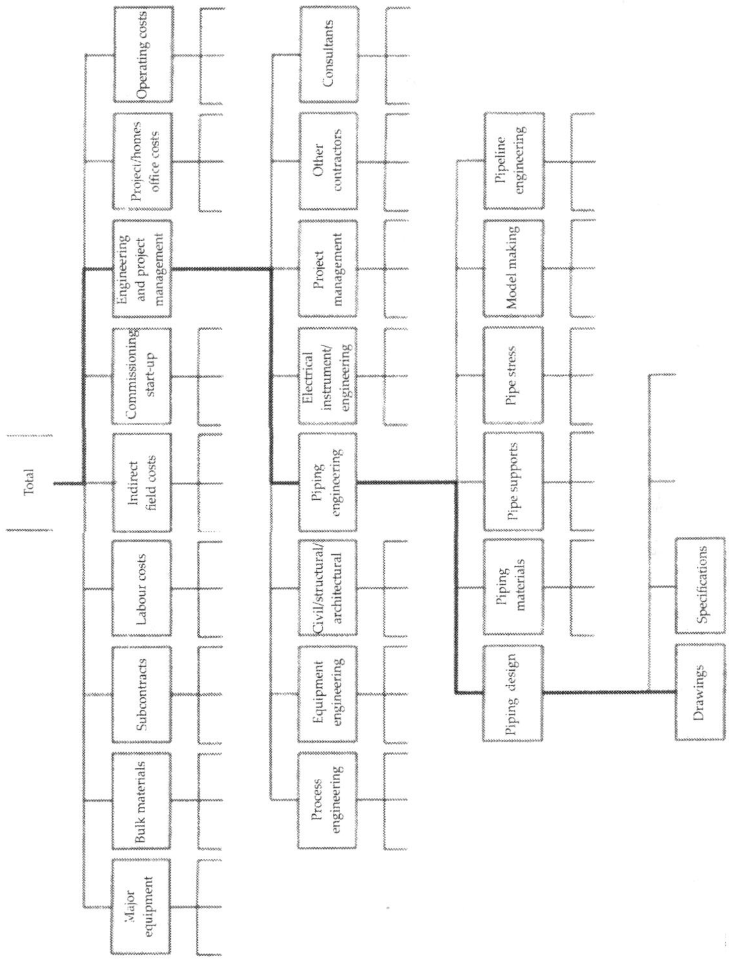

Figure 5 2: A typical code of account (The Joint Development Board sponsored by the Association of Cost Engineers and the Royal Institute of Chartered Surveyors, n.d.)

Framework to help with future project estimates while providing high level cost data for cost control purposes.

While giving itemized data on project costs it is additionally vital that the coding framework ought to be as brief as could be expected under the circumstances but then simple to utilize, implement and understand. The cost code may utilize a progression of four alpha numericals to recognize classifications of cost. Utilizing Figure above and cost code

7421 as an case, cost code 7*** may designate engineering and project administration, 74** pipework, 742" pipework design with the fourth digit designating required deliverables, e.g. drawings, specs, and so forth.

It is fundamental to deliver an effective coding framework ahead of schedule in the project which mirrors the levels of detail required since once the project is running it will turn out to be progressively hard to create itemized cost information underneath that permitted by the first coding framework.

The code of record ought to permit its utilization on a wide range of projects. This won't just encourage precision in cost distribution and announcing yet will over some stretch of time help the estimators by giving good assessing information.

Approval of funds

Before doing spendings on a project it is important to get permission and power/ authority from the proprietor to make spending on behalf of the proprietor. It is typical with extensive projects for this power, frequently named sanction, to have the capacity to be allowed just by the proprietor's governing body.

Techniques to be taken after while applying for sanction power of assets differ between organizations; in any case, it is typical for an avocation document for the project to be set up as a reason for the assessment to continue or forsake. It is likely that a particular project will be contending with various different projects for restricted accessible assets, especially funds, and for each project to be surveyed on its benefits by the top managerial staff, or other high level expert, with an end goal to guarantee that the organization's assets are put into the projects which will give it most extreme future advantage. (The Joint Development Board sponsored by the Association of Cost Engineers and the Royal Institute of Chartered Surveyors, n.d.)

The avocation document will contain all the key data on the project to help this basic leadership process. Key data to be incorporated will typically contain the accompanying:

- Project portrayal;
- Project goals;

- Technical parameters, aspects and factors:
    - Technology and innovation to be utilized;
    - Project area;
    - Size of the facility;
    - Crude material necessities;
- Product yield;
- Business factors:
    - Advertise patterns;
    - Expected advantages (e.g. product deals);
    - Expected activities of contenders;
- Monetary assessment (capital expense versus expected profit for venture):
    - Estimate of project cost;
    - Project plan;
    - Project execution procedure;
    - Appraisal of risks.

Projects travel through a progression of stages from feasibility studies through to appointing and handover. It is typical for the endorsing of project supports additionally to be staged in a comparative design. This guarantees constrained assets are made accessible amid the practicality / feasibility and front end advancement periods of the project and that huge sums are just authorized when adequate data has been created on which to base an educated choice to continue.

Cost control techniques

An extensive variety of strategies are utilized by the cost engineer to recognize potential cost issues on a project. These procedures differ contingent upon the sort of project consumption, and keeping in mind that this distribution does not endeavor to detail these different strategies for control it is helpful to value a few of the all the more generally utilized strategies for approach.

Control of design costs

Design costs, comprising of design, acquirement and project supervisory exercises, can be observed and figured by gathering data on advancement accomplished and worker hours consumed and changed over to cost by applying recorded man-hour rates.

Data accessible to the cost engineer on which to base cost forecasts will regularly include the accompanying:

- Arranged man-hour consumption by genre / discipline
- Arranged advance by genre / discipline
- Original man-hour use by genre / discipline
- Original advancement by genre / discipline
- The genre heads' appraisal of worker hours to finish.

In view of this data the cost engineer can determine the worker hours and in this manner the cost to finish the design exercises, by utilizing the most proper of the accompanying three techniques:

- The worker hours required to finish supposing productivity of 100% in view of the left behind arranged man-hour consumption;
- As technique 1 yet with the outcome balanced by the productivity

$$\text{Productivity} = \frac{\text{Planned man-hours to data}}{\text{Actual man-hours to date}}$$

$$\therefore \text{Total forcast man-hours} = \frac{\text{Total planned man-hours}}{\text{Productivity factor achieved to data}}$$

Figure 5 3: Productivity factors for predictions (The Joint Development Board sponsored by the Association of Cost Engineers and the Royal Institute of Chartered Surveyors, n.d.)

The productivity factor in technique 2 is accomplished by dividing the worker hours foreseen to be utilized as a part of accomplishing a measure of work divided by the original worker hours used in accomplishing an indistinguishable measure of work as appeared in above Figure.

Care ought to be taken while ascertaining the productivity factor to guarantee that an obvious fall in productivity is not caused by changes not yet identified or accepted which have in this way not been consolidated into the present control estimate. When utilizing this determined productivity factor to predict worker hours to finish, thought must be given to factors which may have impacted productivity to date (for instance worker hours consumed on acquaintance) and regardless of whether they will impact future execution. In light of this evaluation it might be important to modify the factor appropriately to get a more practical prediction.

The cost engineer's pronounced prediction of aggregate hours ought to be founded on the rule that amid the beginning times of a project the prediction will be the man-hour difference to date added to the amended control estimate. Between this stage and last phases of design the prediction will be the modified control estimate separated by the productivity factor depicted above. Amid the last phases of design the engineer's prediction of hours to finish will be added to the worker hours consumed as the prediction of aggregate to hours.

By encompassing this appraisal for each design engineering discipline all the time it is conceivable to figure the last man-hour consumption furthermore, in this manner cost of the aggregate design exertion for a project and to distinguish issue areas and potential increases to spending plan.

Control of equipment costs

It is typical for the control estimate to be founded on a sensibly exact equipment numbers delivered amid the early design period of the project. It is likewise typical for the estimator to have acquired spending citations from providers on which to base his appraisal of equipment costs included inside the estimate.

It is essential for the cost engineer to check the legitimacy of these estimates as tenders are gotten for the buying of these different equipment things. Cost increase to the estimate arrangement ought to be quickly recognized and announced, all together that remedial activity can be considered and actualized. Purposes behind increase in spending plan may incorporate yet are not constrained to the accompanying:

- More expansive materials indicated than expected in the financial plan;
- Expanded limit or size - incidentally as a result of overdesign;
- Superfluous necessities composed into the determination;
- Poor choice of bidders;
- Foreign currency vacillations

This disturbance in cost gives some sign as to where to begin searching for the reasons for the problems and it is apparent that restorative move can be made in specific cases. Note that when exploring cost disturbances the probability of a low estimate ought not be ruled out; but rather this ought to be considered simply after all roads of restorative activity have been depleted. Once it has been conceded that the monetary allowance is low all endeavors towards remedial activity will stop.

Control of bulk material costs

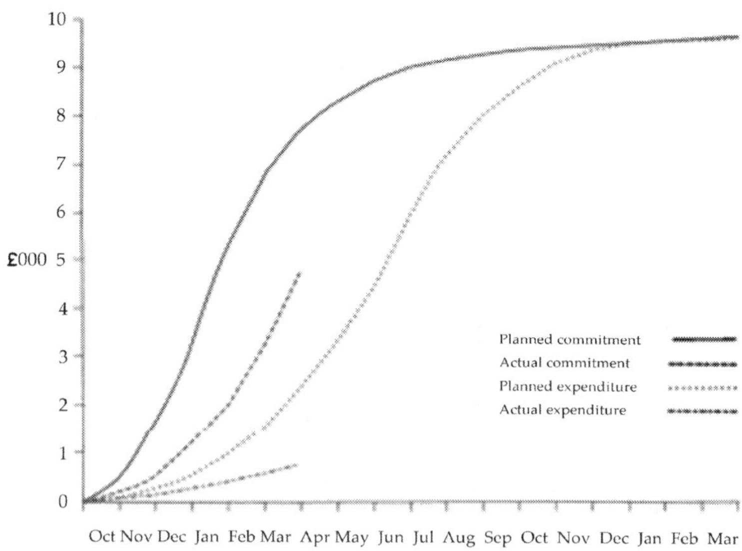

Figure 5 4: Curve of consumption and commitment (The Joint Development Board sponsored by the Association of Cost Engineers and the Royal Institute of Chartered Surveyors, n.d.)

Exception reports

As a method for quickly distinguishing issue regions where advancement is not up to the mark, or man-hour consumption unnecessary, special case reports will be created on an incessant customary premise for the duration of the life of the project.

The special case reports will list those exercises, or individual records, where anticipated culmination dates are later than the action early wrap up date or where an exorbitant man-hour cost change exists.

Cash flow forecasts

An income prediction / forecast will regularly be asked for by the project which will be delivered by successfully evaluating the project plan. This is accomplished by spreading the spending against the periods over which that spending is being brought about in accordance with a suitable consumption / spending profile, so as to create a foreseen income month to month consumption for the rest of the project.

There is frequently a prerequisite for both long-and short interval conjectures and predictions generally covering the ensuing three months. The fleeting prediction would be utilized by the financing team to draw cash down from the proprietor or co-ventures in availability for the installment of month to month solicitations.

DOCUMENT CONTROL

Introduction

The quantity of drawings, particulars, information sheets and endorsements created on a noteworthy project by the proprietor, designer, construction contractual workers, sub-contractual workers, providers and so forth can keep running into a large number, each of which may go through a progression of modifications.

Engineering projects are intricate and include substantial design and construction workforces working in parallel or sequentially on numerous diverse genres. Each genre relies upon data given by others and each thusly gives data to others. Such data may identify with space necessities or to the utilitarian or basic reliance of one framework on another.

This high level of reliance, together with the typical tight timetables and vast workforces implies that inability to deliver a report on time, or inability to remark on a report created by others as per the set time scales can cause severe postponement and or costly change to the design and the construction work.

The advancement of archives, regardless of whether delivered by the proprietor, designer, contractual worker, or provider and at whatever area, should be followed through the different phases of design audit, remark and issue.

Main activities

Productive coordination between all the design controls and providers is crucial. Strict control should be practiced over the advancement of the key deliverable and their accessibility, so as to guarantee that all people requiring the utilization of the data approach the most recent amendments constantly.

On significant projects it is regularly important to set up a record control group to furnish the project with a record administration benefit covering all archive control necessities and related capacities from contract honor to project handover, potentially stretching out into the arrangement of manuals important to the consequent operation and upkeep of the plant.

The principle exercises of the record control group will be to give a data administration to train engineers, construction contractual workers and providers, in different areas covering enrollment, observing, printing, dissemination, assisting and status investigating all reports delivered by the project.

Despite the fact that exemptions might be put forth in the defense of Documentation of a private nature, report control will be the point of convergence for the enrollment, checking and appropriation of project records.

The group will hence:

- Set up an arrangement of report / document referencing
- set up an archive / document following framework to control and provide details regarding the era, issue, receipt, survey and return by each genre;

- Give an arrangement of report / document stockpiling and recovery

Control methods

Document and Report control will work at the level at which singular records can be distinguished and followed, which is normally what might as well be called level 4 of the planning framework. The record control group will be in charge of setting up and keeping up the record database at this level.

Data streams into the project from sub-contractual workers and providers, and so forth in addition flowing out from the design group. With a specific end goal to guarantee that the required data is gotten, in the required arrangement, in the required number and at the required time a report timetable will be issued with each buy request or subcontract, under which there is a design duty, as-manufactured drawing obligation or something like that. Record control will liaise with the originator of the data as to design while the expediters will guarantee receipt of the required archives at the fitting time.

Registration of documents

All documents and archives produced in or for the project, will be recognized by a special number to encourage enrollment, reference and recovery. Enrollment of every deliverable will record at least:

- Archive / document number
- Title of the subject
- Date of first and current issue;
- Correction/status/date;
- Motivation behind issue (e.g. for survey; for design; for enquiry; for construction);
- Document reference.

Transmission of documents

Archives and documents will for the most part be issued under front of a printed version letter, electronic message or other proforma methods for recording its issue.

It is vital to consider the wording of the transmittal which ought to unmistakably express the date of issue and the purpose behind which it is issued. In the event that, as in the instance of drawings, determinations, and so forth being issued to a design or construction contractual worker, it ought to be clarified whether the issue is or is most certainly not a guideline to continue with the work demonstrated subsequently.

It is additionally imperative that the archive controller knows not just that drawings were issued yet additionally that they were gotten through a few means of formal affirmation. To depend on the contractual worker recognizing a crevice in a grouping of transmittal slip numbers is not a practical choice when time is of the embodiment.

Review of design deliverables

Deliverables got from sub-contractual workers and providers requiring survey by others will typically be appropriated to the surveying divisions under front of an inner transmittal that records issue date and the date by which remarks are required, Approval cycles will be followed and recorded and the archive control group will facilitate remarks not gotten by the required date with a specific end goal to keep up the program.

Distribution

As already expressed, it is essential that individuals from the project work provide the most recent data. This is particularly valid on account of the design group.

Plainly not every person on a project requires duplicates of each of the a huge number of reports that a noteworthy project will create, and thusly a record conveyance framework will be built up at a beginning time and ceaselessly kept up to meet the prerequisites of the project, in this way limiting superfluous conveyance.

The matrix will distinguish every one of the reports which it is expected will be created on that particular phase of the work. The archive control group will welcome every individual from the project to recognize on the lattice which of the archives they wish to get as a feature of the standard dissemination. In expansion a refreshed archive rundown will be accessible consistently, either in printed copy frame or on a PC database, against which an engineer can keep an eye on the accessibility of records that he had not initially distinguished as being required to get on the standard dispersion rundown and furthermore to watch that he is presently utilizing the most recent data.

The report control group can guarantee that engineers dependably have access to the most cutting-edge data yet it is up to the individual engineer to guarantee that he is utilizing what is accessible.

As-built drawings

The exercises vital for the readiness of as-fabricated drawings may differ relying upon whether PC aided design (CAD) is being utilized. The contractual worker will be required to increase duplicates of the drawings issued to him for construction purposes, to demonstrate any progressions made at site with a specific end goal to guarantee that on culmination the proprietor has a total and exact record of the work. On the off chance that CAD is utilized the data contained on the increased drawings will for the most part be moved into the PC database. (The Joint Development Board sponsored by the Association of Cost Engineers and the Royal Institute of Chartered Surveyors, n.d.)

Reporting

The status of design deliverables must be known whether project advancement is to be observed. Status audits and reports will along these lines ordinarily be created at what might as well be called planning level 4 utilizing a database.

Before the beginning of engineering design, each genre will survey the number of drawings and different deliverables required to fulfill the finish project scope. This posting will be utilized by the record control group to build up the underlying database against which control will be exercised. Arranged and figure dates will be set by the planning engineers and as every deliverable is gotten by record control, real dates will be held inside a "live" working environment, empowering the present status of expectations to be known and revealed. The database will be refreshed as additional data ends up noticeably accessible.

The report control group should keep up total records of archives connected to a detailing framework, which will ordinarily consider:

- Following of a wide range of deliverable reports with full update history-designer and provider drawings and details, demands,

- Buy orders, and so on.;
- Following progressions and deletions;
- Logging equipment to records;
- Conveying various record numbers for proprietor, contractual worker and provider with variable numbering structures to suit all needs

Retention of Documents

The measure of printed material created on a noteworthy project is monstrous. A few of it is crushed yet many records should be held over critical timeframes.

Diverse customers and distinctive territories will each have specific prerequisites with respect to progressing liabilities which must be set up earlier to the initiation of the project all together that the important reports are held for the essential time frame.

MATERIAL CONTROL

Introduction

The cost of equipment or materials on an engineering project is one of the most huge cost parts in the estimate and the considerable cost of materials neglecting to touch base at the required time or at the wrong area makes postponement and interruption in the construction exercises frequently out of all extent to the estimation of the missing things.

Regardless of whether the material controller is from the project administration group and managing 'free issue' material just, or whether part of a contractual worker's group providing all material, the general necessities of material control render as it is.

The material controller is in charge of guaranteeing that materials are gotten:

- At the set time;
- At the set area;

- In the set amounts;
- To the set specs

The material controller will create and work a framework that screens status against these prerequisites, investigates and supports stockpiling conditions. furthermore, techniques and records any support attempted amid the construction time frame, keeping in mind the end goal to guarantee that material and equipment is in great condition when introduced and that any certification is not negated.

Traceability

Because of the risky type of process plants, notwithstanding the business contemplations expressed above, there will much of the time be outside necessities to fulfill the affirming experts, insurance organizations and statutory specialists with respect to material quality and traceability. Such necessities are keeping in mind the end goal to guarantee that the quality and respectability of the finished plant are to the suitable standard, that the plant can be worked securely, with a negligible risk of failing and that recognizable proof and traceability of supply is encouraged in case of equipment or material failing.

Traceability is typically required for essential things, for example, equipment, piping, fittings, steel support and structural work, and so forth., and records are subsequently typically kept up through a coding structure that recognizes the provider,cluster /batch number and area inside the completed project so that if essential the area of all other comparative parts from a similar group can be built up and the things reviewed or supplanted. (The Joint Development Board sponsored by the Association of Cost Engineers and the Royal Institute of Chartered Surveyors, n.d.)

The methods for giving traceability will rely upon the nature of the material. For instance, electrical links ought to have all the required data imprinted on the link at consistent interims, while each bit of pipework or basic steelwork must have a heat and cast number stamped on it. Numerous proprietors and design engineers now require that such numbers are likewise on the conveyance notes. At the point when a bit of pipe or plate is cut from the first stamped thing of material, the heat and cast numbers must be exchanged on to the piece that is left unnumbered, generally that material won't be permitted to be joined into the completed unit and will be worth scrap only.

Coordination

Amid the acquirement stage the material controller will go about as organizer between the construction contractual worker and the design office, by educating the design engineers of the coding frameworks to be utilized for material control purposes and giving expediters with the needed nearby dates. The material controller will likewise look for the proficient utilization of material by collecting remittances for contingency and waste with all gatherings.

Coordination regularly incorporates 'settling'. Settling is a term utilized for masterminding the different shapes and sizes of steel plate required by the design for the purpose to lessen wastage when they are cut from the standard rectangular plate conveyed from the steel provider. Settling will as a rule be attempted by the manufacture contractual worker and checked by the design engineers. Once "settled" the material controller should record settling through material distribution sheets.

Site storage

Materials and equipment that have been accurately conveyed, will be moved from the accepting territory into site stockpiling. Such capacity differs from a laydown region normally for pipe, steel, and so on., to a stockroom for electrical or instrumentation equipment. Note ought to be taken of any unique prerequisites for security of the materials, particularly on a multi-contractual worker or multi-project site, and for the conservation and assurance of the materials for example, warmed storage. The issue of materials for manufacture and establishment will differ between projects however it is common to demand materials from the stores as they are required to be joined into the work.

Once in the capacity range the materials will be under the quick control of a stores association who either answer to the material controller or work to strategies accepted with the material controller. The stores association will be furnished with the essential lifting and transport plant and different things vital for the sheltered taking care of and assurance of the materials, machinery and equipment.

The material controller is in charge of guaranteeing that the equipment and material are kept in great condition. This will include consistent examination of stores and guaranteeing that equipment is liable to an arrangement of preventive upkeep that may include general review, oil or turning of engines, pumps, and so on.

Precise material control will be accomplished by the keeping of records of receipts, stocks and issues and the cross-referencing of amounts on arrange and conveyance predictions. Along these lines, material lack reports can be created with the goal that incite move can be made to acquire extra material, i.e. by assisting exertion or the situation of new requests.

Especially substantial things of equipment will be required to be conveyed on a foreordained date and off-stacked straightforwardly on to arranged establishments to maintain a strategic distance from twofold handling.

Spares

Material controllers will regularly arrange the interface between the construction contractual workers and the plant operations by recording and giving over surplus materials and spares to the plant operations together with all related affirmation, and so on.

Breakdowns on a manufacturing plant are an insurable risk and the longer a plant is sit without moving the higher the cost to the insurance organization. Subsequently insurance organizations as often as possible require that specific key spares are kept in store with a specific end goal to keep away from delay caused by ordering, making and conveying of new parts. Such spares are frequently alluded to as insurance spares and a material controller must guarantee that such spares are given over to the operations amass in great condition and with the vital documentation.

Close out

A file of test endorsements will be delivered by the material controller for giving over to the proprietor together with an arrangement of as-fabricated drawings and isometrics which contain reference to the heat and cast numbers, and so forth. On project ending, the material control records will be utilized to convey out a material compromise all together that all materials are represented and costed. Surplus material may then be:

- Sold back to the providers (through 'purchase back' provisions in original purchase orders);
- Unloaded to intrigued organizations;
- Sold for scrap worth.

CHAPTER 6

Procurement of Materials and Equipment

INTRODUCTION

This section tends to the acquirement of materials and equipment. Acquirement of construction contractual workers and sub-contractual workers are canvassed in chapter 9.

SCOPE OF PROCUREMENT ACTIVITIES

All through all enterprises there are five directing rule that oversee the acquisition work; they are known as the 'Five Rs':

- Right amount
- Right quality
- Right provider
- Right cost
- Ideal time

To guarantee consistence with these standards, the acquisition work typically involves the exercises identifying with obtaining, speeding up, investigation and delivery. in spite of these exercises is material control, which is considered in Section 5 , Section 4 on material control. The acquirement work is in charge of guaranteeing that all equipment and materials, together with related drawings, reports and information are conveyed in such time as will empower the engineering, construction and authorizing program to be met and consequent plant operations completed. ,

REQUISITIONS

A demand is an announcement of necessity arranged by a client office (typically the engineering or construction office) which is utilized as the principle against which materials and equipment on a projects are obtained. Orders will contain data in regards to extent of supply, amount, particular and last area inside the work, together with subtle elements of any exceptional prerequisites. Required conveyance dates to meet the construction calendar will be connected by the arranging division.

THE PROCUREMENT CYCLE

The term 'acquisition cycle' is oftentimes used to show the time taken from the recognizable proof of a material or equipment necessity to the arrangement of a buy order. The cycle is built up at the arranging phase of a project and conveyance periods evaluated, with the goal that the program of order creation can be planned to meet site conveyance prerequisites.

Drawings and information from a provider might be basic to the design; conveyance dates of the equipment may not accordingly be the main controlling variable. Exercises inside the acquisition cycle are regularly as takes after:

- Receipt of enquiry demand (by acquisition);
- Pre-capability of bidders;
- Issue of enquiry;
- Receipt of tenders;

- Shortlisting of givers;
- Engineering and specialized investigation of tenders;
- Planning of offer rundown;
- Receipt of procurement demand (by acquisition);
- Arrangement of procurement arrange

Time taken for each of the above exercises relies upon the versatile nature of the equipment and materials to be bought, be that as it may, standard circumstances are accepted for arranging purposes.

Standard circumstances will keep running from fruition of the past movement and might be founded on a run of the mill enquiry for a bundled equipment thing. For mass materials, times will be diminished impressively, and for major or complex equipment things, times will increment, particularly when the provider is required to complete design work to consent to delicate prerequisites. The acquirement part on a project begins at the initiation of the point by point arranging stage. The acquirement cycle and individual conveyance periods will be joined into the arranging plan, to guarantee that design data and the materials or equipment things are accessible to meet design and construction prerequisites. This activity not just checks the authority of the general program (design and construction), yet in addition empowers arranging of required obtainment labor to take place. (The Joint Development Board sponsored by the Association of Cost Engineers and the Royal Institute of Chartered Surveyors, n.d.)

THE PROCUREMENT TEAM

For real tasks it is basic to have a devoted acquirement group. The administrator / manager of the group will report straightforwardly to the project director. Other than his acquirement obligations the obtainment administrator will be some portion of the project administration group. All acquirement matters on a project will be managed by the acquisition supervisor and he will liaise straightforwardly with the owner's project director or obtainment facilitator (depending on size of project) regarding all matters concerning the buy of materials and Equipment.

The general duties of the acquirement group are as per the following:

- Obtaining: recognizable proof and assessment of potential providers, enquiry issue, delicate receipt and investigation, arrange situation/administration close-out of the request;
- Assisting: guaranteeing the convenient receipt of design information and materials equipment to meet design/ construction prerequisites;
- Review: guaranteeing that materials equipment requested are to the right particular and meet the required quality;
- Shipment; guaranteeing that materials and equipment discharged by examination are sent in an auspicious and savvy way to the right site; additionally in charge of keeping up import records and contact with custom experts.

COMMUNICATION

The whole relation fashioned by a buy order amongst buyer and provider is based on great correspondence. Where correspondence joins break, blunders happen and strains in this relationship occur. Correspondence is clearly a two-way process and effectiveness will be improved where goodwill and trust are produced. The goal of good correspondence is to maintain a strategic distance from issues, yet in the event that issues do happen, great correspondence should to guarantee that they are recognized early and have less affect on the work. All gatherings to a request ought to be looking to make a fruitful long drag relationship and great correspondence is essential to the accomplishment of this point.

TERMS AND CONDITIONS

Purchase order versus contract

Buy orders are obviously rightfully restricting contracts; many organizations, in any case, allude to the moderately basic documentation required to purchase material as a buy order and the bigger, more mind boggling documentation required to place, control and assess work of design or construction as a contract. In view of the difference in substance and strategy for organization they are every now and again took care of by particular segments. (The Joint Development Board sponsored by the Association of Cost Engineers and the Royal Institute of Chartered Surveyors, n.d.)

Diverse organizations have distinctive translations of what involves a buy order and what involves an agreement. The clearest division between the two is that where work is required at a construction site this should to be the subject of an agreement, with a buy order being appropriate where the course of action is for supply and conveyance as it were. Be that as it may, where there is a vast design component to the work an agreement is habitually utilized as a part of put in of a buy request, to give added security to the buyer. Most organizations will have their own arrangement regarding this matter, yet it should to be noticed that the choice between buy request and contract is once in a while one of value alone.

Obligations of purchaser and supplier

The buyer's essential commitments under a buy order are to characterize clearly what is required and to pay the cost accepted when things are accurately conveyed by those requirements. However, this as expressed is an extremely oversimplified approach and the buyer has different commitments.

Once a request is set, there might be design documentation for the buyer to favor as indicated by approval plans accepted ahead of time with the provider, so delays don't influence conveyance dates. For complex or high worth orders, advance installments might be accepted with the goal that when particular turning points are come to (either time or occasion based or both), installments wind up noticeably due. The buyer additionally resolves to take conveyance at indicated dates either on an ex-works premise (buyer orchestrates gathering) or at a construction site.

For the provider, his commitments can be condensed in three words quality, amount, time. By tolerating a request, the provider consents to give the predefined item(s) to the right determination and quality, in the rectify amount, and to deliver the merchandise in the accepted time scale. Other commitments, for example, arrangement of design information, advance reports, solicitations what's more, support of records likewise exist yet are auxiliary to the fundamental three. Where conveyance is made to a construction site, at that point the stores association or construction/ contractual worker will take conveyance of the things. They will be in charge of off stacking and safe stockpiling of things, until required for construction. All together that installments can be made to providers, they will guarantee that things conveyed are right in number and that no deficiencies or harm have happened in travel.

Basis of order

Buy requests will specify what is required, to what particular, the amount required, the cost and conveyance date and area. The buy order is affirmation of every one of that has gone between the buyer and provider from a straightforward phone call toward one side of the range, to a few periods of definite arrangements at the flip side. An assortment of reports might be either referenced in, or connected to, the request and these extend the request's substance. Such records will go from advisers for illuminating prerequisites, determinations, and so forth, to the accepted terms and conditions under which the request is set. The request from a buyer speaks to acknowledgment of a provider's offer and without accepted conditions; an enforceable contract may not exist. Such conditions are very important to the two gatherings since they guarantee that duties are known.

General terms and conditions

Most organizations have their own particular buy order terms and conditions what not would want to exchange utilizing their own particular terms as opposed to those of another organization. Be that as it may, it is uncommon in the engineering business to acknowledge a provider's terms and conditions even where close imposing business model circumstances happen.

In this manner, an imperative piece of the acquisition procedure is the understanding of terms and conditions before arrange position. Where the contractual worker is acting in the owner's name or on a 'for and for benefit of premise then it is common for owner's terms and conditions to be utilized. However a contractual worker undertaking acquirement all alone sake will as often as possible use 'in-house' conditions 'with amendments to reproduce any special terms of the main contract. (The Joint Development Board sponsored by the Association of Cost Engineers and the Royal Institute of Chartered Surveyors, n.d.)

From a provider's perspective, the essential terms of any request identify with guarantees, liabilities, punishments and terms of installment. It is regular in the oil what's more, gas industry for maintenance agreements to be looked for, because of the time it takes for undertakings to be finished, with a guarantee time of three years not being abnormal. As to of installment, the standard is as often as possible a private installment, yet point of reference installments might be connected to finish of drawings or accommodation of information where such is required to advance the principle single standard determination set up furthermore, authoritative costs diminished. Such requests can be put on a for each project premise or for a particular time, covering an owner's necessities on a few projects.

Call-off orders

Call-off orders or supply agreements are used to envelop bulk material requirements where final quantities are not known at the date of original order placement

When issuing enquiries on this premise, it is essential to have a sensibly exact measure of aggregate quantities and things required, with the goal that best cost and rebates can be acquired. From a provider's viewpoint, such orders are advantageous as they help generation arranging and secure a long pull responsibility. Cancel orders are utilized for channeling materials, auxiliary steel, valves, link, and so forth. For orders that reach out more than one year, and where costs are expanding, an acceleration equation can be consented to share the dangers between the gatherings.

Terms and conditions for such requests would not change to any grand degree over those for a chance requirement, however would deliver things extraordinary to cancel circumstances, i.e. no certification of conclusive amounts, value acceleration formulae, purchase back game plans, and so on.

Bonds and guarantees

A variety of business apparatuses are accessible to a buyer when drafting terms and conditions, which are utilized both to give solace to the purchaser against potential issues and to restrain dangers to the purchaser.

Those most generally utilized are as per the following:

- Parent organization guarantee issued by a definitive parent of a provider ensuring that its backup organization will play out the work as per the request and that if their auxiliary organization should default, that the parent organization will finish or pay for an outsider to finish the work. This sort of parent organization guarantee is promptly given by providers as it does not cost them anything (barring minor organization costs).
- Bank guarantee, issued by the provider's bank to cover progress or advance installments as well as early arrival of maintenance cash held. Such certifications force a monetary risk on the provider that ordinarily diminishes the farthest point of any credit accessible to the provider. Maintenance, withheld from solicitations paid (typically 10%). Arrival of maintenance is as a rule on conveyance (half) with the adjustment paid toward the end of the guarantee time frame.
- An execution bond is issued by a bank or insurance agency for a certain rate of the request worth (normally 10%) which is brought in the occasion of non-execution by the provider.

In choosing which of the above devices (assuming any) to utilize, the purchaser must survey the potential dangers of the provider's inability to perform, against the extra expenses brought about by their utilization. The arrangement of any of the above is definitely not in a provider's enthusiasm as they cost cash to get, defer some portion of his installment or lessen his capacity to get credit. The other most normal technique for lessening the purchaser's presentation to chance is the presentation of exchanged harms. Exchanged harms are typically composed into buy requests to give insurance against late conveyance; they are regularly connected on a raising premise, commonly at a rate of 1% of aggregate request worth every week, up to a most extreme of 10%.

Although generally utilized as a part of procurement orders, sold harms are from time to time conjured, and as a general rule, the potential danger of inconvenience guarantees that providers perform. In the present atmosphere of made strides provider relations, joining forces, and so on. such dangers are viewed as a feature of the recorded antagonistic way to deal with providers and their utilization is much decreased.

Warranties

Most if not all work will be liable to a guarantee (some of the time called a ensure) against blemished workmanship or materials. The time of the guarantee will be set in the different contracts and buy orders and will rely upon the desires of the owner.

Guarantee periods more often than not keep running for a time of a year from the date on which the entire establishment is given over. Since the construction of a plant can take numerous years it takes after that guarantee periods for construction work finished or equipment provided amid the beginning times must have long guarantee periods.

If a primary contractual worker buys equipment, and so on in his possess name at that point, subject to the specific contract conditions, any guarantee gave by that primary contractual worker will incorporate any materials, equipment, subcontract work, and so forth., secured under that agreement. Assuming, be that as it may, a thing of equipment, not bought in the owner's name, is required to have a guarantee in abundance of that applying to the principle contract, unless the primary contract makes arrangement for this extra guarantee prerequisite, it will be important to guarantee for a reasonable course to be set up through which the advantage of the guarantee can be passed from the buyer to the owner either by task of the guarantee or by method for a guarantee.

CERTIFICATION AND CERTIFYING AUTHORITIES

To guarantee that the right materials and equipment are conveyed, it is typical practice to expect providers to give duplicates of material testaments, sort test declarations, and so on, to cover the things requested. Certain documentation might be required with the goal that the last finished plant can acquire the essential endorsements to begin creation. For UK seaward activities, this includes the work of a guaranteeing expert. The affirming expert is a free organization t the plant acquires the authentication of wellness from the Branch of Exchange and Industry. (Salvendy, n.d.) Hat guarantees that the design, materials and equipment have been delivered to the required levels with the goal that

MATERIAL TRACEABILITY

Material certificates, sort test authentications and so forth will be of constrained worth if the material or equipment to which they apply can't be distinguished in the construction site store or in the finished project. United to the need for confirmation from providers is the should have the capacity to look after material traceability. With mass materials, for example, steel plate or pipe, reference numbers likening to those set apart on the applicable confirmation can be 'hard stamped' on the things. Such markings are known as warmth numbers and are rehashed when the material is cut into little pieces. For equipment things, label numbers will be fastened either on nameplates or to metal labels. Both strategies permit simplicity of traceability once introduced so that if there is disappointment or different issues, the first provider can be recognized and the source of material followed.

SPARES AND SPECIAL TOOLS

Spares

There are three important sorts of spares: startup, insurance and working spares.

Startup spares are generally acquired either in the meantime as the principle equipment or independently amid the assembling stage, however regularly on a similar request. Regularly these spares are things that may require on time substitution amid authorizing, i.e. gaskets, globules, channels, jolts, and so forth. Which are of low worth. Contingent upon establishment fundamentals, these spares are conveyed either with the fundamental equipment, or quickly before startup happens.

Protection spares are, as their name proposes, acquired to give protection against an important issue with a bit of equipment. They are ordinarily bought where the cost of a noteworthy breakdown would at the startup time an interruption in yield, along these lines causing genuine financial bad luck. Sorts of spares in this classification are extra rotors for primary oil line pumps, spares for stage cranes and extra turbines for either compressor or power generation bundles. Regularly, such things have protracted lead times and are frequently of high worth. Likewise with charging spares, such things are either acquired at an indistinguishable time from the principle equipment (yet with a later conveyance date) or on an individual buy order.

Working spares are at first for it is possible that maybe a couple of years' operation; these are customarily bought independently from the principle equipment and regularly are bought by the owner direct, instead of by the project acquirement group. Requests might be put later than the primary equipment, with conveyances occurring on time before plant startup. Starting working spares are just acquired direct from the equipment provider for the beginning necessity, for promote progressing needs things are purchased coordinate from the producers.

The cost of a wide range of spares ought to be mulled over when making the first provider choice, as high working expenses can exceed any underlying preferred standpoint on introductory capital cost. Because of the high cost of spares, both in capital speculation and capacity costs, providers are currently being induced to store spares at their cost, as opposed to expect the owner to buy and store.

Special tools

A few things of equipment require the utilization of uncommon instruments for the motivations behind establishment, charging or support. The necessity for uncommon devices will be distinguished inside the order and joined in the enquiry.

Supplier data

Equipment supply bundles may have an extensive design content, and the appropriate receiving of information (drawings, determinations, methodology, affirmation, and so forth.) to ongoing design reasons for existing is basic. Now and again, the receiving of such information can be more basic to the general project program than the receiving of the equipment itself. When issuing an enquiry, the purchaser will indicate the information required and the time scale for its receiving. Dates for accommodation will be accepted with the provider before arrange situation, and the provider will be facilitated against these dates. When information are gotten, the designers will be sped up to guarantee remarks or endorsements are made inside the accepted periods (regularly 10-15 days) and any resubmissions required from the provider will likewise be sped up. Receiving of provider information must be continually checked all through the extend program to keep up design advance and individual provider programs. Last installments should not be made until all information and records are gotten and authorized.

QUALITY ASSURANCE

All requests set are liable to a QA review if so required. Progressively, more providers are met all requirements to BS 5750 or IS0 9000 and are consequently freely reviewed against this standard. Quality confirmation reviews take all things considered one to two days. They cover the provider's full acquirement cycle from receiving of a request to consummation of all liabilities including arrangement of spares. Any issue zones found are followed up later.

PRE-QUALIFICATION OF SUPPLIERS

One of the primary moves to be made at the beginning of a project is the generation of a provider's rundown. The provider's rundown might be delivered from existing organization records including owner assignments and inclinations, or from a project particular rundown. Where an organization list exists, it is as often as possible the case that some type of choice process will be important to diminish the number of potential bidders by a pre-capability survey. The prequalification survey is a straightforward technique for delivering offer records and will as a rule be the sole technique utilized.

The pre-capability survey can either be of a general sort covering all out project prerequisite, or be item particular; it is issued with the target of delivering an offer list(s) of organizations adequate to guarantee sufficient competition, and including those companies most equipped for giving the products and no more conservative cost, while keeping up quality and calendar prerequisites. For high worth or complex things, it might be important to supplement the pre-capability process with interviews and for visits to coming up providers. (Bessant, n.d.)

Reference should be made to chapter 9 covering contractual worker sub-contractual worker determination for the important necessities in defining the delicate rundown, assessment, review necessities, and so on where these end up plainly obligatory under the EU orders.

Evaluation

The pre-capability survey tends to general organization points of interest, fund, involvement, references, abilities, assets, QA, security, and so on., what's more, each inquiry can be scored or weighted as required. For assessment of reactions, either a target or subjective strategy can be utilized. With an target technique each inquiry is scored (typically out of 10) and a weighting given to show relative importance. Of the three principal areas to be evaluated, focuses to be available are as per the following:

- Monetary: most recent three years' records, brokers, parent organization subtle elements;
- Technical: item range, past activities, design and testing offices;
- Assets: producing fine elements, area, sorts of staff, equipment, and so on.

TENDER LIST

Because of the assessment of the pre-capability survey, an accepted tender will be determined. The quantity of givers required relies upon many-sided quality of things required and enquiry worth. For most things, the tender will fluctuate in the vicinity of three and six organizations.

ENQUIRIES

Enquiry substance will change dependent upon general project necessities and the sort and estimation of materials or equipment to be acquired. Nonetheless, for a commonplace expansive size project, an enquiry will more often than not comprise of the accompanying documentation.

Technical

- Acquirement: items and amounts required;
- Information prerequisites:
- Information required, accommodation periods, amounts, introduction technique, and so on;
- Specs;

- Packing necessities;
- QA necessities;
- Assessment levels

Commercial

- Covering letter: return date, conveyance required, and so on;
- Directions to Tenderers: accommodation strategy, duplicates required, points of interest to be secured;
- Type of tender: valuing points of interest;
- Terms and conditions

Once issued, any inquiries from givers identifying with the enquiry will be steered to the appropriate purchaser, who is in charge of their provoke reply. Any responses to inquiries that influence each of the givers will be issued as an enquiry release and issued to every one of the givers. Solicitations for augmentation to delicate return dates might be allowed if programs stay unaffected and where a few givers are making comparable solicitations. Where enquiries have been issued on a 'fixed delicate' premise then all such tenders will be opened in the meantime and an underlying record of costs made. The level of fixed tenders differs between organizations however most tenders with an worth more than £50000 will be liable to this technique. For low worth enquiries, tenders can now and then be acknowledged by fax to speed arrange situation. (The Joint Development Board sponsored by the Association of Cost Engineers and the Royal Institute of Chartered Surveyors, n.d.)

Once opened, duplicates of tenders are circulated by the purchaser to the different masters for their specialized investigation that proceeds in parallel with the business assessment. Where a few organizations have submitted tenders, it is common to wait list maybe a couple organizations at a beginning time, to accelerate the determination procedure.

When this has been refined by method for an underlying investigation and acceptance with the delicate evaluation group, the purchaser will focus his investigation on the shortlisted tenders.

Analysis

On fulfillment of the short rundown, the purchaser and designer will create isolate point by point investigations of the tenders. Any questions with connection to the tenders will be routed to the offering organizations and answers will be incorporated into the fundamental examination. Despite the fact that the business and specialized examinations are created independently, a joint suggestion is made on fulfillment, i.e. the best business offer that is actually adequate. Separated from looking into the underlying capital cost the business examination will likewise address the accompanying things:

- Value legitimacy;
- Conveyance period;
- Spares;
- Capacity Limit;
- Money related strength;
- Subordinate expenses.

Despite the over, any capabilities to terms and conditions, guarantee period, terms of installment and so forth will be considered and the impact on the buy order assessed.

Where vital, meeting(s) will be held with givers to clear up both business and specialized parts of offers, with the goal that an exhaustive recommendation to honor can be made. The purchaser will mastermind and seat such gatherings and will guarantee that minutes of gatherings are delivered and accepted with the givers. Regularly givers will be required to erase any of their standard supply conditions which may have been incorporated into tenders. (Alistair Barros, n.d.)

Award

Once the investigation has been affirmed, the purchaser will set up a tender document what's more, buy order for endorsement and mark. The request will mirror the costings contained in the investigation and any specialized assentions came to resulting to enquiry issue. Requests are marked and issued once the important buy demand has been gotten from the architect. In the event that a pre-grant meeting is vital, this will again be masterminded by the purchaser what's more, understandings achieved reflected in the genuine buy order. Two duplicates of the request will be sent to the provider, with one duplicate going about as the authority arrange affirmation to be marked and come back to the purchaser.

Any change of a business or specialized nature to equipment or materials after a request has been put, will be secured by the issue of an official revision to arrange. The purchaser will regularly hold duty for the organization of the request through to arrange finish off.

EXPEDITING AND INSPECTION

Expediting

On the off chance that providers and buyers completed the greater part of their accepted commitments on time, there would be no requirement for the expediting function. Be that as it may, this is regularly not the case and speeding up is utilized to advance providers and keeps a extend mindful of potential conveyance issues. Speeding up is embraced to suit the program fundamentals of a project and is worried about not just physical conveyance of equipment and materials, yet additionally the stream of design information. For projects with broad design fundamentals, the requirement for drawings and information from providers is of vital significance to keep up general project plans. Along these lines, before honor of a request, the purchaser will accept archive accommodation dates with the provider what's more, these dates will shape the premise of any required expediting effort.

The sort of facilitating to be utilized will shift from visits to the provider's location to consistent contact by phone or fax, or a blend of both. For visits, the expediter will create a point by point report demonstrating documentation status and request advance. Issue territories will be featured, medicinal activities proposed (if vital) and a figure of conveyance made. Report status will likewise be refreshed with most recent guaranteed dates for accommodation. Assisting conjectures are encouraged again into the project program so that issue ranges can be instantly recognized. Not all slippages in conveyance dates cause issues, and zones that influence programs are concentrated on. As a supplement to coordinate assisting, providers are frequently asked for to submit consistent (regularly month to month) advance reports.

Recurrence of speeding up visits and request advance is audited on a general foundation with the goal that its need and adequacy can be measured and needs looked into. Where essential, ordinary assisting visits can be supplemented by either administration gatherings or visits, to strengthen the speeding up process.

The speeding up work is additionally in charge of guaranteeing that equipment furthermore, materials are conveyed in the right amount and to the correct area. The expediter goes about as the contact point between the provider and the conveyance site, and admonishes destinations what materials and equipment things are being conveyed and when. Nothing moves from a provider's works without the expediter's earlier learning and authorization.

Inspection / inspection

Investigation can be viewed as an industry confirmation benefit in that its capacity is to guarantee that what has been requested is in actuality what is provided and conveyed. Ideally such an administration would not be required and things conveyed would be to the right details and quality; sadly this is not the case and investigation is required to "police" the fabricate of obtained materials and equipment.

Investigators will commonly have an assembling or designing foundation and specific down to earth and specialized aptitudes and capabilities. It might be important to utilize outsider inspectorates to attempt examination in the interest of a buyer when the thing to be assessed is of a expert nature, or because of the topographical area of the maker.

Investigation builds up agreeableness of materials and equipment through consistence with design, ruinous and non-dangerous testing and weight, confirmation, stack, fracture, execution, useful and operational testing rehearses upheld by significant guaranteeing articulations.

Investigation is required not exclusively to fulfill the prerequisites of the buy order yet in addition to guarantee consistence with statutory controls. The level of investigation to be connected relies upon the idea of the necessity and on the many-sided quality of the equipment and the criticality of the thing to the project. Hence, mass materials are generally subject just to a last assessment, while a compressor bundle will be reviewed at different stages all through the assembling and testing process. To accomplish this an assembling quality control design is created by the provider for approval, which cautions the provider to the level of examination required amid the life of the request what's more, which contains witness, hold and investigation focuses at which review is required before proceeding with produce or conveyance.

Examination by the buyer should not be permitted to dissolve the provider's obligation to guarantee that what is produced is to the required quality furthermore, determination. Reports are created after every assessment visit to feature issue zones and to fill in as an advance beware of the request. On conclusion of construct, the equipment will be examined against last affirmed documentation and if satisfactory, an examination discharge note will be issued permitting arrangement for shipment. In the event that issues exist, a 'punch-list' of things requiring rectification or culmination will be created by the investigator, also, the provider will be required to clear this rundown before reinspection.

As a feature of his exercises, the reviewer will check all fitting accreditation and documentation and guarantee that finished information dossiers go along with extend necessities.

SHIPPING, TRANSPORT AND INSURANCE

Shipping and transport

On most projects, providers are in charge of the conveyance of equipment furthermore, materials to site. Once an examination discharge note has been issued, the expediter will acquire points of interest from the provider of weights, sizes and amounts and will issue exact delivery directions. Subtle elements will be encouraged to site to caution the site administration of approaching conveyances. While in travel, the provider stays in charge of the equipment and orchestrates vital protections.

For vast seaward activities or abroad undertakings, the administrations of a cargo sending organization are regularly used to control and execute the development of equipment and materials. The utilization of a forwarder builds the control a extend has nearby conveyances, lessens the cost as a result of mass shipments and enables needs for developments to be set, so pressing site necessities can be quickly tended to. At each phase in the conveyance procedure, the project knows about status and real area of the equipment and materials. Arrangement ahead of time of anomalous burdens happens with the goal that courses are distinguished and vital licenses got. It is normal for the forwarder to give an in-house organizer to complete this work and communicate with the extend group. Cargo reserve funds are recognized by looking at the forwarder's cost with that of the provider's for conveyances. (Horngren, n.d.)

Whichever strategy for conveyance is utilized, the buy request will plainly state conveyance terms in light of the universally perceived Incoterms terminology. For conveyances by method for cargo forwarder, requests will be put on the premise of 'ex-works stacked onto buyer's vehicle'.

Packing

Any exceptional pressing, bundling and marking necessities will be secured in the buy order and will assess site conditions, stockpiling, and so on. For seaward activities, send out pressing is regularly determined so the equipment can be put away securely for a while. Unless pressing necessities are determined by the buyer the provider's standard is typically acknowledged.

Customs and Excise

All imports into the UK from the EU are liable to VAT, while import obligations may apply to products from outside the EU. Where equipment or materials are designated for seaward undertakings, the buyer can recover VAT paid and is absolved from import obligations through internal handling help (IPR). As a result of the 'single market' inside the EU, points of interest of EU imports presently must be given by the buyer on a month to month premise to the traditions experts. Where a cargo forwarder is utilized, it is normal to incorporate all traditions records and contact in their degree.

Dispatch of goods

When materials and equipment are discharged for dispatch, the expediter will On time either the site immediate or through a material controller, that things are accessible for conveyance and he will acquire points of interest of weight, estimate, and so on in addition to assessed time of entry. On receiving at site, things will be checked and reviewed to guarantee that the sum total of what things have been gotten, that they are as per the expressed prerequisites and that no harm has happened in travel. Aside from checking the genuine products, printed material will likewise be checked to guarantee that the important accreditation, documentation and review discharge notes have been gotten.

On the off chance that the things are adequate, a report will be issued affirming that the material has been gotten (some of the time called a material got report (MRR) or merchandise got note (GRN)), which will permit any vital installments to be made. In the event that the things are inaccurate or harmed, or if there are missing things, and so on then a report (now and again called a harm disparity report or DDR) will be issued by the material controller to the purchaser for activity. The purchaser will talk about any vital activity with the provider what's more, installments will be deferred or diminished until the point that the issue has been cleared.

It is vital on account of harm being supported, or products being lost, that the material protection methodology is promptly enacted.

PROGRESS MEASUREMENT AND REPORTING

Estimation of obtainment exertion or advance is not a correct science and is refined in an assortment of courses, none of which is absolutely attractive. The accompanying techniques are as of now being used:

- Weighting things in the acquisition cycle, for example, enquiry issue; arrange arrangement, and so on and after that allotting all out accepted worker hours in like manner. Advance is then measured against these breakthroughs for the aggregate obtainment exertion;
- Setting general points of reference for prudent exercises, for example, dates for all enquiries issued, all significant

equipment conveyed, and so on. Despite the fact that this gives characterized focuses to acquirement exertion, it makes no recompense for genuine products of common sense of accomplishment;

- Month to month focuses for accomplishment of exercises, for example, arrange issue;
- Evaluating worker hours utilized on a month to month premise against accepted manhour add up to.

For reports, the acquirement group creates singular request reports (facilitating in addition to assessment), general status writes about requests set in addition to pre-arrange organize and a project arrange enlist. Reports can be customized to suit singular project fundamentals with the goal that the total obtainment action is overseen. (Bessant, n.d.)

CHAPTER 7

Quality Assurance and Quality Control

INTRODUCTION

In the present business world, it has turned out to be increasingly important to obtain a supposed and acknowledged standard of measured quality before major clients will consider doing business with an organization. This genuinely late quality upset has gained pace and validity from the beginning of the BSI Kitemark, through BS 5750 to the present IS0 9000 series and is now the subject of internationally concurred standards covering products and enterprises.

The requirement for a quality system standard was supposed and created during the Second World War and subsequently in the US space program. It was clear that items and systems purchased from contractual workers would not be accessible on time or be good and work to specification unless each worked according to a firmly defined and documented quality system.

In Europe the fundamentals of an appropriate system relevant to defense contractual workers were standardized through NATO as Allied Quality Assurance Publications (AQAPs), and were initially distributed in the UK by the Ministry of Defense as defense standards (DEF Stans.). These standards were subsequently utilized by the British Standards Institute as a basis for the introduction in 1979 of the BS 5750 quality system standards.

This procedure of advancement finished in the publication of IS0 9000 standards by the International Standards Organization, ISO, in 1987. On their publication the original British standards were pulled back and supplanted by the BS 5750: 1987 series as the UK form of the IS0 9000 reports.

Quality management applies to all items and administrations - fields as assorted as generation engineering, nourishment, materials, building and development, vitality, wellbeing and social administrations, transport, pharmaceuticals, nearby government, bookkeeping and law. Those in charge of quality in every one of these fields, apply logical and specialized abilities and specific management systems to accomplish the standard of item or administration required by the client at the correct cost and the ideal time. All professionals, whether accountants, architects, engineers, surveyors or solicitors, now have the opportunity to exhibit their proficiency and consistency through the utilization of a demonstrated quality management system within their own businesses.

DIMENSIONS OF QUALITY

Eight dimensions of product quality management can be used at a strategic level to analyze quality characteristics. The idea was characterized by David A. Garvin. A portion of the dimensions are mutually refreshing, whereas others are not—change in one may be to the detriment of others. (Alistair Barros, n.d.) Understanding the trade-offs fancied by customers among these dimensions can help construct an upper hand. Garvin's eight dimensions can be summarized as takes after:

- Performance: Performance refers to a product's primary operating characteristics. This measurement of quality includes measurable attributes; brands can usually be ranked impartially on individual aspects of performance.
- Features: Features are additional characteristics that enhance the appeal of the product or administration to the client.
- Reliability: Reliability is the probability that a product won't

fail inside a particular day and age. This is a key component for clients who require the product to work without fail.

- Conformance: Conformance is the accuracy with which the product or administration meets the predefined standards.
- Durability: Durability measures the length of a product's life. At the point when the product can be renovated, estimating permanence is more intricate. The thing will be utilized until the point when it is not any more cost-effective to work it. This occurs when the repair rate and the related costs increase meaningfully.
- Serviceability: Serviceability is the speed with which the product can be put into benefit when it breaks down, as well as the skill and the behavior of the administration individual.
- Aesthetics: Esthetics is the subjective measurement indicating the sort of reaction a client has to a product. It speaks to the individual's personal inclination.
- Perceived Quality: Perceived Quality is the quality attributed to a decent or administration based on backhanded measures.

Performance

Performance indicates to a product's primary operating characteristics. This measurement of quality includes measurable attributes, so brands can usually be ranked equitably on individual aspects of performance. Overall performance rankings, in any case, are harder to grow, especially when they include benefits that few out of every odd customer needs. Performance is often a source of dispute amongst customers and suppliers, particularly when deliverable are not adequately characterized inside specifications. The performance of a product often impacts the profitability or reputation of the end-client. As such, many contracts or specifications incorporate damages related to inadequate performance. The topic of whether performance contrasts are quality contrasts may rely upon circumstantial predispositions yet predispositions based on functional prerequisites, not taste. Some performance standards are based on subjective inclinations, yet the inclinations are universal to the point that they have the constrain of a goal standard. (Bessant, n.d.)

Features

Features are additional characteristics that enhance the demand of the product or administration to the client.

Similar intuition can be applied to features, second dimensions of quality that is often a secondary aspect of performance. Features are the "fancy odds and ends" of products and administrations, those characteristics that supplement their basic working. Examples incorporate free beverages on a plane, permanent-squeeze cycles on a washing machine, and automatic tuners on a shading TV. The line splitting primary performance characteristics from secondary features is often hard to draw.

Reliability

Reliability is the probability that a product won't fail inside a particular era. This is a key component for clients who require the product to work without fail.

This measurement reflects the probability of a product malfunctioning or failing inside a predetermined day and age. Among the most well-known measures of reliability are the mean time to first failure, the mean time amongst failures, and the failure rate per unit time. Because these measures require a product to be being used for a predefined period, they are more relevant to durable merchandise than to products and administrations that are devoured instantly.

Reliability normally turns out to be more important to buyers as downtime and maintenance turn out to be more costly. Farmers, for example, are especially sensitive to downtime among the short crop season. Reliable gear can mean the distinction between a decent year and ruined harvests. Yet, customers on different markets are more attuned than any other time in recent memory to product reliability as well. PCs and replicating machines certainly compare on this basis. (Waldemar Karwowski, n.d.)

Reliability may be firmly related to performance. For instance, a product specification may characterize parameters for up-time, or acceptable failure rates. Reliability is a major supporter of brand or company image, and is viewed as a fundamental measurement of quality by most end-clients. I.E., late market research demonstrates that, especially for ladies, reliability has turned into an automobile's most sought attribute.

Conformance

The result of two example procedures to demonstrate the meaning of the two approaches to conformance

The measurement of conformance explains to what degree a product's outline and operating characteristics meet well-known standards. This measurement owes the most to the traditional approaches to quality spearheaded by specialists like Juran.

All products and administrations include specifications or something to that affect. At the point when products are created, these specifications are set and a target is set, for instance the materials used or the measurement of the product. The target as well as the tolerance (the range of allowed deviation from the target) is characterized. One issue with this approach is that there is little interest for whether the specifications have been met exactly as long as far as possible is met.

From one viewpoint, this can lead to the supposed "tolerance stack-up". Whenever at least two parts are to be fit together, the extent of their tolerances often decide how well they will match. Should one part fall at a lower point of confinement of its specification and a matching part at its upper breaking point, a tight fit is doubtful. The connection is probably going to wear more rapidly than one made from parts whose dimensions have been focused all the more exactly.

This issue can be addressed by taking an alternate approach to measuring quality. Instead of measuring a basic conformance to specifications, how much parts or products separate from the ideal target is measured. Utilizing this approach, process 1 (see picture) is better despite the fact that a few things fall past specification limits. The traditional approach would have favored process 2 because it delivers more things inside as far as possible. It was demonstrated that the issue of "tolerance stack-up" is more shocking when the dimensions of parts are more distant from the target than when they group around it, regardless of the possibility that a few parts fall outside the tolerance. This approach requires a crisp take a gander at the regular procedure quality factor of 'defect rate', to take into account the fact that two parts may each pass the 'tolerance test' separately however be unusable when the attempt is made to consolidate them. (George A. Steiner, n.d.)

In profit organizations, measures of conformance normally focus on precision and opportuneness and incorporate checks of get ready bumbles, unexpected deferrals and other standard errors.

Durability

Durability measures the length of an item's life. Exactly when the item can be settled, assessing durability is more confounded. The thing will be utilized until the point that the moment that it is no savvier to work it. This happens when the repair rate and the related costs increment essentially. In fact, durability can be portrayed as the measure of utilization one gets from an item before it deteriorates. After such a large number of hours of use, the fiber of a light expends and the globule must be supplanted. Repair is abnormal. Budgetary investigators call such items "one-hoss shays" (Oliver Wendel Holmes number).

This way to deal with durability has two important deductions. To begin with, it suggests that durability and unwavering quality are solidly related. An item that frequently falls flat is most likely going to be rejected sooner than one that is more solid; repair costs will be reliably higher and the buy of a dangerous brand will look essentially more attractive. Second, this approach induces that durability figures should be deciphered with mind. An expansion in item life may not be the eventual outcome of specialized changes or the usage of longer-lived materials. Or maybe, the concealed monetary condition just may have changed.

Serviceability

Serviceability joins the buyer's simplicity of acquiring repair advantage (illustration: access to profit centers or potentially simplicity of self-advantage), the openness of organization faculty (case: simplicity of getting an arrangement, ability of repair work force to tune in to the customer), and the dependability of organization (case: paying little mind to whether the organization is performed right the main gone through). Wellness and simplicity of repair is the speed with which the item can be put into advantage when it separates, and the ability and the behavior of the organization work force. (Salvendy, n.d.)

Purchasers are worried about an item separating and in addition about the time before advantage is restored, the favorability with which advantage arrangement are kept, the nature of dealings with advantage work force, and the repeat with which advantage calls or repairs neglect to revise exceptional issues. In those situations where issues are not quickly settled and grumblings are reported, an organization's protestation taking care of strategies are additionally prone to influence customer's definitive assessment of item and organization quality.

Some of these factors reflect shifting individual gauges of satisfactory organization. Others can be measured indifferently. Clients may stay disappointed even after finish of repairs. How these protestations are taken care of is important to an organization's notoriety for quality and organization. Inevitably, benefit is most likely going to be influenced also. Organizations differ extensively in their ways to deal with grievance taking care of and in the importance they join to this component of serviceability. Some do their best to decide dissensions; others use lawful creations, the calm treatment and comparative ploys to repulse disappointed clients.

For instance starting late, General Electric, Procter and Gamble and diverse organizations have endeavored to proper buyer disappointment by introducing without toll phone hot lines to their customer relations offices.

Important qualities for serviceability estimation are: advantage guarantee, parts guarantee, parts accessibility, number of sensible separation to merchant profit centers, separation to profit parts center merchant, separation to profit parts center individual, length of sit tight for advantage arrangement, date-book of preventive upkeep, delegates tune in to clients, information in regards to repairs, obliging organization centers, repaired precisely first time, advantage time in respect to various merchants, guarantee claims took care of without contention, normal repair cost/year, benefit contract, underestimation of organization cost and course of action of credit auto.

Feel or Style

The elegant properties of an item add to the character of an organization or a brand. Blames or imperfections in an item that decrease its classy properties, even those that don't reduce or adjust diverse measurements of quality, are frequently causing for expulsion.

Style alludes to how the item looks feels, sounds, tastes, or smells. It is obviously a matter of individual judgment and an impression of individual aura. Everything considered, there give off an impression of being a couple of examples in customers' rankings of items on the premise of taste.

A present examination of quality in 33 sustenance categories, for instance, found that high capacity was every now and again related with qualities, for instance, "rich and full flavor, tastes normal, tastes new, extraordinary fragrance, appealing looks. Esthetics likewise insinuates the "outside" feel of the item. (Salvendy, n.d.)

The style estimation contrasts from subjective criteria relating to "performance" in that elegant choices are not almost general. Not all people lean toward "rich and full" flavor or even concur on what that implies. Organizations along these lines need to look for a forte. On this estimation of quality, it is hard to please everyone.

Perceived Quality

Perceived is not generally reality.

Purchasers don't generally have complete information about an item's or organization's traits; meandering measures might be their reason for looking at brands.

An item's durability for instance, can on occasion be observed particularly; it for the most part ought to be translated from different unmistakable and elusive parts of the item. In such conditions, pictures, publicizing and brand names - inferences about quality instead of the truth itself-can be basic. Subsequently, both Honda - which makes autos in Marysville, Ohio-and Sony - which fabricates shading TVs in San Diego-have been hesitant to advance that their items are "made in America".

Notoriety is the essential stuff of perceived quality. Its vitality originates from an implicit relationship: that the quality of items today is like the quality of results of yesterday, or the quality of stock in another product offering is like the quality of an organization's built up items.

QUALITY ASSURANCE

There are numerous "definitions" of these two words, which being subjective often prompt confusion at first experience. Be that as it may, as the peruse gains more trust in the subject, an example of comparative interpretations will emerge. The most cited definition is that given in BS 4778:

Every one of those planned and systematic actions important to give sufficient certainty that an item or administration will fulfill gives basics for quality. As each agreement for administration or item may shift from the past one, the view of quality can promptly be believed to rest with the client. On the off chance that every client is fulfilled, then quality has been accomplished.

With the goal that consumer loyalty is not an 'erratic' or random occasion, it takes after that a systematic approach is required. The quick development of a specific system, for example, BS 5750 (now renamed IS0 9000) is because of the way that it offers a framework of exercises that assess and measure the installed quality system, and accommodates it to be recorded and re-measured for consistency and keeps an eye on corrective actions and (it is to be trusted) changes. (George A. Steiner, n.d.)

It is quality records, especially in the form of review reports, that allow management to audit the adequacy of their business practices, and which give management and their clients the knowledge that their items or administrations are 'quality assured'.

TOTAL QUALITY MANAGEMENT

Total quality management is something the majority of organizations are pursuing today in some form. An attentive is, as Professor Steven Wearne (UMIST) stated, 'Quality can lose reason, turn into a paper work out, onerous and bureaucratic'. Each opportunity to counteract such a happening and to construct a quality culture is required.

Another UK standard, BS 7850, which assesses total quality management, is now accessible. The target of a total quality management system is to distinguish every one of those issues important to guarantee the accomplishment of the composed destinations.

THE STANDARDS

The model for a system of quality management is IS0 9000 that can be embraced by organizations and independently assessed. It adequately combines the disciplines and methods that make the accomplishment of reliable quality conceivable. The standard additionally concentrates consideration on the quality system, gives more powerful documentation and, when legitimately comprehended and connected, makes quality change inevitable. This makes the executed system more viable in reducing errors, thus saving cash and improving consumer loyalty.

IS0 9000 and IS0 9004

IS0 9000-1 gives guidelines to the choice and utilization of the standards and explains the structure of the standards. It defines the quality ideas and clears up the distinctions and interrelationships between them. Guidelines are likewise accommodated the determination and utilization of the quality system standard that can be utilized for internal quality management or outer quality assurance purposes. The importance of this area of the standard is that it explains when and how to utilize IS0 9001,9 002 or 9003 when there is an agreement between a purchaser and a provider. IS0 9004-1 is a Guide to Quality Management and Quality System Elements which is composed as a guide to the introduction of internal quality management systems. (Bessant, n.d.)

ISO 9001

IS0 9001 is the Specification for Design/Development, Production, Installation and Servicing. It applies to those organizations and organizations that are involved in the original outline, improvement, generation and subsequent operation of an item or administration as determined by the purchaser. Providers secured by IS0 9001 include configuration gatherings and PC installers.

ISO 9002

IS0 9002 is the Specification for Production and Installation. It applies to organizations or organizations providing a set up item or administration to a concurred specification. The provider is here assumed to have no critical part in new outline or in the servicing, which in many cases are under the control of the purchaser. This prompts more restricted duties regarding management, internal auditing and training. Providers secured include makers and providers of weight vessels and valves.

ISO 9003

IS0 9003 is the Model for Quality Assurance in Final Inspection and Test. It applies to organizations where conformance of the item to determined necessities is required to be certainly shown by inspection and tests on the finished item.

THE SYSTEMS

The introduction of IS0 9000 quality systems alone does not ensure item or administration quality. This has to be refined through a combination of good systems and specialized capability. IS0 9000 sets a framework for the trip of continuous change towards total business greatness.

Quality plans

Quality planning is a basic piece of any quality system and the quality arrangement depicts how and when the quality prerequisites will be executed.

A quality plan is a report, or a few archives, that together specify quality standards, hones, assets, specifications, and the succession of exercises important to a specific item, administration, project, or contract. Quality designs ought to characterize:

- Objectives to be accomplished (for instance, attributes or specifications, uniformity, adequacy, feel, process duration, cost, regular assets, usage, yield, constancy, et cetera)
- Steps in the processes that constitute the working practice or methodology of the organization
- Allocation of duties, authority, and assets amid the distinctive periods of the process or project
- Specific archived standards, practices, methodology, and guidelines to be connected

- Suitable testing, investigation, examination, and audit programs at fitting stages
- A recorded technique for changes and alterations to a quality arrangement as a process is moved forward
- A technique for measuring the accomplishment of the quality objectives
- Other activities important to meet the objectives

At the most irregular amount, quality objectives and plans ought to be incorporated with general key designs of the organization. As organizational objectives and plans are conveyed all through the organization, each capacity forms its own most ideal path for adding to the best level objectives and objectives. (Salvendy, n.d.)

At bring down levels, the quality arrangement expect the part of a significant arrangement. Such designs may take a wide range of forms relying upon the result they are to deliver. Quality designs may likewise be spoken to by more than one kind of record to deliver a given result.

A case of this is an assembling organization that machines metal parts. Its quality arrangement comprises of pertinent systems (portraying the generation process and obligations), relevant workmanship standards, the estimation resistances adequate, the depiction of the material standards, et cetera. These may all be separate records.

More factor information that relates to a specific client might be explained on singular work orders (now and then called voyagers). Work orders specify the machine setups and resistances, operations to be performed, tests, assessments, handling, storing, bundling, and conveyance ventures to be taken after.

A working level quality arrangement interprets the client necessities (the what) into activities required to deliver the coveted result (the how) and couples this with pertinent techniques, standards, practices, and conventions to specify precisely what is required, who will do it, and how it will be finished. A control design may specify item resistances, testing parameters, and acknowledgment criteria. While the phrasing may vary, the essential approach is comparative for benefit and different sorts of organizations.

Deployment and Documentation of Quality Plans

Quality plans result from both carried vital quality policies (which are connected to organizational key plans) and from the specific lawful regulations, industry standards, organization policies and methodology, inside rules, and great practices expected to meet clients' necessities for items or administrations.

Key level quality plans are produced and conveyed through the vital arranging process. These wide based quality plans turn into the rule for each function's or office's supporting quality arrangement. Where fitting, each function or office may create and inside convey working level quality plans.

Working level quality plans often are the subsequent document(s) from a creation planning function. All things considered, this documentation often incorporates outlines, a duplicate of the client's order, references to appropriate standards, practices, systems, and work guidelines, and points of interest on the most proficient method to deliver the specific item or administration.

At the point when the item or administration is created, the arranging reports might be increased by investigation documentation, SPC graphs, and duplicates of delivery archives and client required confirmations. In the process, the plans are transformed from reports to records. In a completely automated system, the reports specified may well be intuitive PC screens, got to at operators' workplaces and control focuses. These screens internally move toward becoming records when operators, inspectors, shippers, and others make PC passages to the screens. (Salvendy, n.d.)

A finished arrangement of networks, created by a quality function deployment (QFD) process, may satisfy a part of an organization's quality arrangement. The purpose of QFD is to catch and convey the clients' needs and necessities all through the organization.

Reporting the quality plan(s) has different utilizations, for example,

- Assuring conformance to client necessities
- Assuring conformance to outside and inner standards and methods

- Facilitating traceability
- Providing objective proof
- Furnishing a reason for preparing
- Together with various plans for the organization's items, administrations, and projects, giving a premise to assessing the adequacy and efficiency of the quality management system.

Cost of Quality The cost of quality is the entirety of costs a project will spend to forestall poor quality and some other expenses acquired because of yields of poor quality. Poor quality is the waste, errors, or inability to address partner issues and project prerequisites. The expenses of poor quality can be separated into the three categories of aversion, evaluation, and disappointment costs:

- Prevention costs: These are decided costs an organization brings about to guarantee that errors are not made at any phase amid the conveyance process of that item or administration to a beneficiary. Cases of avoidance costs incorporate quality arranging costs, instruction and preparing costs, quality organization staff costs, process control costs, statistical surveying costs, field testing expenses, and preventive support costs. The cost of anticipating botches are constantly a great deal not as much as the expenses of assessment and correction.
- Appraisal costs: These incorporate the expenses of confirming, checking, or assessing an item or administration amid the conveyance process. Cases of examination costs incorporate getting or approaching assessment costs, interior creation audit costs, test and review costs, instrument upkeep costs, process estimation and control costs, provider assessment expenses, and audit report costs.
- Failure costs: A project brings about these expenses in light of the fact that the item or administration did not meet the prerequisites and must be settled or supplanted, or the administration must be rehashed.

Typical contents of a quality standards system

Most quality systems address the following 20 points. Some ark plain as day, all are to be found in IS0 9000 and in most broad messages regarding the matter of quality:

- Management responsibility and organization;
- Quality system or program;
- Contract review;
- Design control;
- Document control;
- Purchasing;
- Purchaser supplied items;
- Product identification and traceability;
- Process control;
- Inspection (testing and surveillance);
- Inspection (measuring and test equipment);
- Inspection (test and operating status);
- Control of non-conforming products;
- Corrective action;
- Handling, storage, packing and delivery;
- Quality records;
- Quality audits;
- Training;
- Servicing;
- Statistical techniques.

ISO 9000 gives the design to quality systems. It is communicated in wording that can be connected to a wide range of extensive, medium and little undertakings, in designing and constructing the item or providing the support of the prerequisites of the purchaser.

The current international (ISO) standard is the ISO 9000 series, and the outstanding British quality standard, BS 5750, was renamed as BS EN ISO 9000 in 1994, to line up with the numbering for easy international acknowledgment. The content was and is identical, and future updates will be correspondingly in step.

Quality manual

The most irregular amount of the corporate quality plan is the quality manual that formally expresses the organization's arrangement on quality, and gives a synopsis of the governing principles in the main areas of the organization's business. A common quality plan will distinguish a listing of organization procedures with a synopsis depiction of substance to indicate linkages and connections between the different procedures. This might be appeared as a stream outline.

The manual will apply to all operations of the organization, and any exercises performed out with the necessities of the quality manual will require prior composed endorsement from those so authorized.

Quality plans

Separate quality plans will be delivered and issued for each of the main areas of the organization's operations and on major undertakings the segments of the quality plan might be included in the venture execution plan. The quality plans will acknowledge the principles lay out in the quality manual and will include the procedures concurred and endorsed by the corporate body relating to particular quality practices, assets and exercises significant to a specific operation. Any takeoffs from the prerequisites of these procedures must have prior composed authorization.

PROCEDURES

Modem management thinking is that work forms should be comprehended and particularly engineered to help the accomplishment of goals.

Records on how choices are come to are thought by a few people to be an issue. Trying to find out what has been considered and what has not, and the reasons for a specific choice, can demonstrate troublesome; however an understanding of the principles followed in reaching a choice is fundamental to validating a quality outline and later to operational basic leadership.

The quality assurance team's goal ought to be to determine the ideal number of procedures. Procedures tend to go substantially further than is entirely compulsory on the basis that such further detail helps productivity. In many cases such solution is counter to effectiveness, is severe and limits initiative and improvement.

Procedures ought to be basic, include just basic material and should just be compulsory for those issues which, if a certain system were not taken after, would represent a genuine danger of failing to meet purchasers' supplies, be a hazard to safe working, cause financial misfortune or rupture lawful prerequisites. It ought to be clarified in the methodology what is compulsory and what is just direction and therefore not compulsory. (George A. Steiner, n.d.)

According to the principle of 'proprietorship', draft procedures ought to be reviewed before final endorsement and issue by the people will's identity in charge of implementing them.

Beneath the level of formal procedures there will be work instructions or strategy explanations that define the subtle elements of how the procedures are to be actualized. Such records ought to be delivered by the individual in charge of the capacity, who should likewise endorse any critical takeoffs from their necessities.

TEAM QUALITY

In spite of the fact that the different owners on a venture are in charge of the usage and maintenance of quality systems within their areas of responsibility the responsibility for quality lays with everybody on a venture.

It is basic to appoint a QA engineer to go about as a point of convergence and impetus for QA matters who will normally be in charge of the following:

Providing an administration and counsel on quality issues;

- Ensuring the usage of fundamental quality systems throughout the venture and any issues not particularly assigned in this area;
- Arranging and where fitting, conducting quality audits throughout the venture;
- Coordinating and standardizing all supplier surveillance exercises.

Contractual workers have in many cases instigated quality chambers and training to involve everybody.

The quality assurance team should expect to "engineer" the work or choice procedures utilized, rather than develop 'a sticky plaster approach' to putting issues great happen.

SUB-CONTRACTUAL WORKER AND SUPPLIER QUALITY

It is now normal to depend on the quality assurance of the sub-contractual worker or supplier rather than on the conventional approach of the purchaser inspecting the work. In certain cases (a typical case being British Steel), the importance of the material is to such an extent that independent inspection is as yet required. In other cases however it ought to be satisfactory to approve that the supplier has sufficient inspection and depend on that.

Suppliers ought to be enlisted to IS0 9001 or IS0 9002, or have identical procedures set up.

A team comprising senior chiefs will often visit immeasurably important sub-contractual workers and suppliers prior to or shortly after honor of an agreement. This is to urge the sub-contractual worker or supplier the importance of their work to the venture, the requirement for quality and convenient finish.

Quality assurance projects will be produced for all obtained items as a feature of preliminary engineering. A performance test will be performed wherever practicable on all suppliers' equipment.

Suppliers will by and large be required to maintain material and test records for inspection throughout the operating existence of the venture.

QUALITY AUDIT

Quality audit is the process of systematic examination of a quality system completed by an inner or outside quality auditor or an audit group. It is an important piece of organization's quality management system and is a key component in the ISO quality system standard, ISO 9001.

Quality audits are ordinarily performed at predefined time interims and guarantee that the establishment has obviously characterized inside system monitoring strategies connected to compelling activity. This can help decide whether the organization consents to the characterized quality system processes and can include procedural or comes about based evaluation criteria. (Alistair Barros, n.d.)

With the update of the ISO9000 procedure of standards from the 1994 to 2008 procedure, the concentration of the audits has moved from simply procedural adherence towards estimation of the genuine adequacy of the Quality Management System (QMS) and the outcomes that have been accomplished through the execution of a QMS.

Audits are an essential management instrument to be applied for checking objective proof of processes, to survey how effectively processes have been realized, for judging the adequacy of accomplishing any characterized target levels, to give prove concerning depletion and end of issue zones. For the advantage of the organization, quality auditing ought report non-conformances and corrective activities, as well as feature zones of good practice. Along these lines different divisions may share information and change their working practices subsequently, likewise adding to continual improvement.

Quality audits can be an essential piece of consistence or regulatory prerequisites. One case is the US Food and Drug Administration, which needs quality auditing to be achieved as a article of its Quality System Regulation (QSR) for medicinal devices.

A few nations have received quality audits in their advanced education system (New Zealand, Australia, Sweden, Finland, Norway and USA) Initiated in the UK, the process of quality audit in the instruction system concentrated fundamentally on procedural issues as opposed to on the outcomes or the efficiency of a quality system execution.

Audits can likewise be used for security purposes. Evans and Parker (2008) portray auditing as a standout amongst the most capable security monitoring strategies and 'a successful approach to stay away from complacency and feature gradually deteriorating conditions', especially when the auditing centers not just around consistence but rather viability.

The processes and errands that a quality audit includes can be overseen using a wide variety of programming and self-evaluation apparatuses. Some of these relate specifically to quality regarding wellness for reason and conformance to standards, while others identify with Quality expenses or, more precisely, to the Cost of poor quality. In dissecting quality costs, a cost of quality audit can be connected over any organization instead of just too regular creation or get together processes.

QUALITY IMPROVEMENT

There are always two bases for quality improvement: one is the responsive basis, where issues found during audits or other surveillance techniques are recorded, and preventive measures taken to guarantee no repeat. The second is the proactive, which is more troublesome as improvement suggests knowing the starting point. There is, therefore, a need always to counsel information gathered with the intent of examination for quality improvement, or information that are accessible that can be utilized for quality improvement investigation, the proactive being less very much defined.

The main instrument for recognizing, in a proactive way, the requirement for quality improvement, is the generation, where conceivable, of stream graphs for each working method. The sensible succession can be seen and, therefore, the normal result can be assessed.

Continual improvement process

A continual improvement process, additionally regularly called a constant improvement process (contracted as CIP or CI), is a continuous effort to enhance items, administrations, or processes. These efforts can look for "incremental" improvement after some time or "achievement" improvement at the same time. Delivery (client valued) processes are continually assessed and enhanced in the light of their efficiency, viability and adaptability.

Some consider CIPs to be a meta-process for most management systems, (for example, business process management, quality management, project management, and program management). W. Edwards Deming, a pioneer of the field, considered it to be a piece of the "system" whereby feedback from the process and client were assessed against organizational objectives. The way that it can be known as a management process does not imply that it should be executed by 'management'; but instead only that it settles on decisions about the usage of the conveyance process and the outline of the conveyance process itself. (Waldemar Karwowski, n.d.)

A more extensive definition is that of the Institute of Quality Assurance who characterized "nonstop improvement as a progressive ceaseless change which seems to be: '... concentrated on expanding the adequacy as well as efficiency of an organization to satisfy its arrangement and objectives. It is not restricted to quality activities. Improvement in business procedure, business comes about, client, worker and provider connections can be subject to continual improvement. Put basically, it signifies 'improving constantly'.

Kaizen

Some effective executions use the approach known as kaizen (the interpretation of kai ("change") zen ("great") is "improvement"). This strategy ended up noticeably renowned from Imai's 1986 book Kaizen: The Key to Japan's Competitive Success.

- Feedback: The core principle of CIP is the (self) impression of processes.
- Efficiency: The motivation behind CIP is the recognizable proof, lessening, and end of problematic processes.
- Evolution: The accentuation of CIP is on incremental, continual strides as opposed to monster jumps.

Key elements of kaizen include:

- Improvements depend on numerous little changes as opposed to the radical changes that may emerge from Research and Development
- As the thoughts originate from the workers themselves, they are

more averse to be drastically unique, and therefore simpler to execute

- Small improvements are less inclined to require major capital venture than major process changes
- The thoughts originate from the gifts of the current workforce, rather than utilizing examination, experts or hardware – any of which could be exceptionally costly
- All workers ought to continually be looking for approaches to enhance their own particular performance
- It urges workers to take possession for their work, and can help reinforce group working, in this manner enhancing worker inspiration.

The components above are the more strategic components of CIP. The more key components incorporate deciding how to build the estimation of the conveyance process yield to the client (viability) and how much adaptability is profitable in the process to address evolving issues.

Six Sigma

Six Sigma (6σ) is an organization of procedures and instruments for process improvement. It was presented by engineers Bill Smith and Mikel J Harrywhile working at Motorola in 1986. Jack Welch made it vital to his business procedure at General Electric in 1995.

It tries to enhance the quality of the yield of a process by recognizing and expelling the reasons for surrenders and limiting variation in assembling and business processes. It exploits an procedure of quality management strategies, basically experimental, measurable techniques, and makes a special framework of individuals inside the organization who are specialists in these strategies. Every Six Sigma project did inside an organization takes after a characterized succession of steps and has specific esteem focuses, for instance: lessen process duration, decrease contamination, and reduce costs, increment consumer loyalty, and increment benefits. The term Six Sigma (promoted in light of the fact that it was composed that way when enlisted as a Motorola trademark on

December 28, 1993) originated from wording associated with measurable signifying the assembling processes. The development of an assembling process can be described by a sigma rating demonstrating its yield or the rate of deformity free items it makes. A six sigma process is one in which 99.99966% of all opportunities to create some component of a section are measurably expected that would be free of deformities (3.4 flawed elements for every million opportunities). Motorola set an objective of "six sigma" for the majority of its assembling operations, and this objective turned into a by-word for the management and designing practices used to accomplish it. (George A. Steiner, n.d.)

Doctrine

Six Sigma doctrines attest:

- Continuous efforts to achieve steady and unsurprising process comes about (e.g. by reducing process variety) are of crucial importance to business achievement.
- Manufacturing and business processes have attributes that can be characterized, measured, and broke down, enhanced, and controlled.
- Achieving maintained quality improvement requires duty from the whole organization, especially from top-level management.

Elements that set Six Sigma apart from past quality-improvement activities include:

- A clear concentrate on accomplishing quantifiable and quantifiable financial comes back from any Six Sigma project.
- An expanded accentuation on solid and energetic management administration and support.
- A clear sense of duty regarding settling on decisions on the premise of certain information and factual techniques, instead of presumptions and guesswork.

The expression "six sigma" originates from measurements and is used as a part of factual quality control, which measures process ability. Originally, it alluded to the capacity of assembling processes to deliver a high proportion of yield inside specification. Processes that work with "six sigma quality" over the short term are accepted to deliver long haul deformity levels beneath 3.4 imperfections for each million opportunities

(DPMO). Six Sigma's implicit objective is to enhance all processes, however not to the 3.4 DPMO level essentially. Organizations need to decide a proper sigma level for each of their most important processes and endeavor to accomplish these. Because of this objective, it is officeholder on management of the organization to prioritize regions of improvement.

"Six Sigma" was enlisted June 11, 1991 as U.S. Administration Mark 1,647,704. In 2005 Motorola credited over US$17 billion in investment funds to Six Sigma.

Other early adopters of Six Sigma incorporate Honeywell and General Electric, where Jack Welch presented the method.[6] By the late 1990s, around 66% of the Fortune 500organizations had started Six Sigma activities with the point of reducing costs and enhancing quality.

Lately, a few experts have joined Six Sigma thoughts with lean assembling to make an approach named Lean Six Sigma. The Lean Six Sigma system sees lean assembling, which tends to process stream and waste issues, and Six Sigma, with its attention on variety and plan, as reciprocal disciplines went for advancing "business and operational excellence". Companies, for example, GE, Accenture, Verizon, GENPACT, and IBM utilize Lean Six Sigma to concentrate transformation efforts on efficiency as well as on development. It fills in as an establishment for advancement all through the organization, from assembling and programming improvement to deals and administration conveyance capacities.

The International Organization for Standardization (ISO) has distributed in 2011 the primary standard "ISO 13053:2011" characterizing a Six Sigma process.[10] Other "standards" are made for the most part by colleges or organizations that have supposed first-party affirmation programs for Six Sigma.

THE WAY FORWARD

The future accomplishment of every one of our businesses is in our hands, to win work, to finish it to the consumer loyalty's, to urge further orders and to guarantee continuing improvement in business techniques. In this scan for improvement the three "soft" management issues of quality, security and condition will rank similarly in importance to the customary "difficult" issues of cost and timetable.

It is plainly obvious that a quality occupation will be a protected employment, and noting that BS7750 for natural systems is based on the old BS 5750, we have the basis for extensive cooperative energy by combining the three subjects into one structure. Purchasers can't renounce their lawful duties to the contractual worker; the law now makes the "business" dependable. Future work must be won by competing in all the above areas, and it will be the most proficient team that wins the business.

CHAPTER 8

Coordination of Procedures and Value Management

INTRODUCTION

The expert organizations related with the UK civil engineering enterprises have created standard reports documents and prerequisites of training which permit any individual taking a shot at any project to have a sensible thought of what his inverse part on another project will be doing.

By examination the engineering enterprises have created from a number of various enterprises and geological zones. In spite of the fact that the basics of organization might be comparable in the different divisions of the industry the way in which it is drilled and by whom can be significantly unique.

Most proprietors, contractual workers and sub-contractual workers are presently subject to a quality framework which requires the organization to keep up procedures covering its exercises, however such procedures may not be completely perfect with those of the proprietor/ contractual worker. With a specific end goal to guarantee productive working it is fundamental to set down those regular exercises important to adjust existing procedures and to set up new procedures at the authoritative interfaces.

The absence of any institutionalization in documentation and working practices implies that there is a risk of irregularity in the procedures. For instance installment necessities may in a few occurrences be expressed in the states of contract, while in others they are incorporated into the remuneration area of the contract, on the other hand they might be integrated in an arrangement of coordination procedures bound into the agreement or accepted ensuing to the honor of the agreement.

Procedures can be separated into administration, design and site procedures be that as it may, the many-sided quality and detail of the documentation will rely upon the nature of the project and all the more critically the prerequisites of the states of agreement.

MANAGEMENT PROCEDURES

All through the entire of this part, remarks have been made with respect to a chain of command of control and reporting being accessible at different levels to suit the necessities of the numerous individuals from the project group and their corporate experts. Accentuation has been set all through on the perplexing nature of engineering projects both as far as design and administration.

The foundation of a solitary arrangement of procedures which covers the by and large organization logic is fundamental for complex multidiscipline projects to work effectively.

The accompanying subjects will typically require to be facilitated to guarantee similarity between procedures created by different associations:

- Interchanges including addresses for correspondence, documenting frameworks, verbal interchanges and electronic mail;
- Engineering matters including prerequisites for record dissemination, registers, drawing and detail arrange, drawing audit and endorsement, as-assembled drawings, affirmation records and working manuals;
- Acquisition, material control and subcontracting, including necessities for buy orders, endorsement of each of the givers/

contractual workers,

- Format drawings of stockrooms and capacity territories, material receipt and control, declarations and traceability and transfer of overflow material;
- staff control;
- QA/QC necessities including quality designs and procedures, assessment records and dimensional control;
- Safety design and safety manual;
- Advancement reporting including prerequisites for revealing, day by day, week after week, month to month, and the arrangement of advancement pictures;
- Planning and advancement control prerequisites including programs at different levels of detail, advancement control and estimation frameworks, histograms and advancement "S" curves;
- Cost control and revealing prerequisites including work breakdown structures, code of accounts, checking and controlling costs, reporting and predicting costs and income plans;
- Invoicing and installment including applications, accommodation, questioned solicitations, installment;
- Contract guidelines/corrections/variations including notice and assessment;
- Insurance claims reporting and assessment;
- Project culmination including last records, acknowledgment endorsements and last documentation.

DESIGN PROCEDURES

The design procedures are likely a standout amongst the most complex arrangement of procedures on any project. Not exclusively are the design procedures inside a solitary design group complex, the endorsement cycles and coordination with other parts of the design procedure, e.g. equipment producers and proprietors, make it fundamental that all groups are attempting to perfect procedures.

Design procedures cover the administration of the design and ought to incorporate the control both of composed data as details and reports, and so on and of drawn data.

In the underlying periods of a project it is fundamental to set up the structure of the design procedures that will be appropriate for every one of the stages of the project. Be that as it may it is not important to have a full arrangement of procedures toward the beginning of a project to cover the later components of the design process, for instance the procedures to cover the survey and endorsement of providers' drawings, such procedures can be issued at a more proper time. It is basic that the procedures are organized to empower them to be integrated at a later date and created to address the issues of the project.

Procedures are typically in view of the proprietor's necessities and created to meet the prerequisites of the project. A proprietor who frequently obtains projects will frequently have point by point corporate predictions which must be conformed to. These can cover the arrangement of drawing data in electronic format reasonable for the proprietor's own particular framework, and at the other outrageous down to the thickness of lines on drawings.

As a project grows so will the design procedures and the proprietor's underlying prerequisites will be extended by the designer's, contractual workers, sub-contractual workers' and providers' design procedures. It is basic that these procedures are adjusted so as to maintain a strategic distance from conflicts and to lessen the risk of confusing frameworks and configurations. One of the key procedures is that covering the status of design data as it is imperative that the issue of all data is controlled all together that the beneficiary of the data comprehends the status of the data and what they should or shouldn't do with it.

SITE PROCEDURES

Site procedures are an arrangement of procedures used to control the work did on the area of site. Procedures cover various subjects, some of which are legitimate to production of site procedures basic to the legitimate operation of a site, especially in the zone of health, safety, while others are basic to the proficient working of the site and organization of the agreements.

Site procedures control different exercises covering:

- Access to the equipment, material and work;
- Work and shift hours;
- Legally binding issues;
- Security/classification;
- Health and safety;
- Advertising;
- Meeting and reporting prerequisites;
- Administration structures;
- Quality confirmation;
- Environmental imperatives (clamor, tidy, and so on.);
- Testing and start-up

The procedures change from project to project and inside a project from contract to contract. Site procedures are regularly in view of the proprietor's prerequisites, which will characterize a progression of confinements and regulatory prerequisites for the project. These are produced by the design group and might be such that they must be incorporated into the delicate documentation to the degree that they influence the delicate cost. Contractual workers may later create and develop these procedures as might be proper for their own control purposes.

PRODUCTION OF SITE PROCEDURES

Site procedures should be set up before the beginning of any site work or related off-site work. The procedures ought to be clear and well composed keeping in mind the end goal to help the proficient running of an agreement. In the event that they are, for instance, inadequately organized and drafted utilizing diverse styles they tend to prompt perplexity and disappointment.

In their drafting it is vital to arrange components, for instance the working hours on location may be confined by different prerequisites, for example, the proprietor directs all work amid set hours, commotion confinements at specific circumstances of day set down in a before planning endorsement, accessibility of the proprietor's site security staff to allow access to the site, and so forth. (Bessant, n.d.) It is normal for a contractual worker to issue site procedures to the project administrator for input or endorsement. This should guarantee that the site will be legitimately controlled and that any limitations, and so on are unmistakably informed to the site group.

The important site procedures ought to be issued under QA procedures as a controlled report to the proprietor, project administrator and contractual workers and through the contractual workers to their sub-contractual workers, and so on.

TYPICAL SITE PROCEDURES

The accompanying is a constrained number of issues that site procedures cover.

Health and safety

Health and safety is an imperative component in site procedures and can ordinarily incorporate licenses to work, site departure and so on.

Instructions

The issue of giving directions is a typical authoritative issue. Site organization procedures should hope to institutionalize the authoritative position and cover matters which are not particularly secured by the agreement.

Ordinarily contracts require that all guidelines might be in composing. However what happens if the contractual worker gets a verbal direction? Should he act on it? It is regular in this case for the contractual worker to affirm the verbal guideline in composing and request a formal direction in composing. It must be clear either inside the agreement or in site procedures whether the first verbal guideline must be followed up on or not before any composed affirmation is gotten and whether inability to dismiss the verbal guideline implies that the direction is affirmed. Comparative issues emerge over the status of site meeting minutes and correspondence to the contractual worker from the proprietor, and so on. (Waldemar Karwowski, n.d.)

Along with the issuing of guidelines most legally binding necessities express that only named people may approve a guideline.

Site procedures must be set up keeping in mind the end goal to effectively control the above.

Drawings and specifications

Drawings and specs are clearly basic for construction to be done. This data must be exact and current; subsequently great site organization procedures are fundamental with a specific end goal to guarantee this is accomplished.

It is common for site organization to have registers of drawings and details including arrangements of superseded data. The control of this data can be extremely mind boggling when both the proprietor and the contractual worker have components of design duty. The proprietor's design may change furthermore, this change needs controlling and conveying. Contractual worker's design regularly requires the proprietor's endorsement, which is typically required inside an indicated day and age. Any disappointment by the proprietor to remark upon or support the record, may in specific conditions be taken as endorsement and any tardy remarks understood as varieties. The circumstance progresses toward becoming more intricate when the proprietor's progressions influence the contractual worker's design and thusly providers' and sub-contractual workers' design. This must be controlled and recorded all together that the work is precisely and effectively completed while guaranteeing any execution criteria, required by the proprietor for the works, are met.

Site records

Site records are various and are normally kept by the contractual worker with duplicates being sent to the agreement organization group. These may cover:

- Climate conditions-temperature least, most extreme, wind, rain, and so on.;
- Assets on site - number of administration work force and agents counting genre and name;
- Plant nearby sort, duty / capacity;

- Health and safety records-hours worked, events, and so forth.

Each of the above might be recorded at various circumstances ranging from hourly to month to month. A few records might be accessible for investigation by the proprietor, some replicated to the proprietor and others agreed with the proprietor.

It is vital that the site records prerequisites are unmistakably distinguished and set up as it is basic to consider to them on occasion of legally binding or quality debate.

Payments

As cash is the spirit of all organizations it is fundamental to set out the procedure by which installments are accepted and made. The procedures ought to be founded on the necessities of the agreement and should cover:

- Timing-month to month, towards the finish of the month or event based;
- Level of help detail required to be submitted with the payment application;
- Individual and address to whom the application of payment is to be submitted, and any necessity to get a receipt on submitting it;
- Observing the advancement of the payment including applications for enthusiasm on late payment

As cash is the spirit of a contractual worker's business this is regularly seen as a key procedure for all gatherings to an agreement.

Insurances

Despite the fact that the checking for arrangement of insurances by contractual worker / sub-contractual workers might be the subject of an administration procedure, site procedures are typically limited to the way in which insurance claims ought to be taken care of.

Arrangement of insurances may incorporate getting insurance in joint names, verifying of the approaches by the proprietor, showing insurance declarations, yearly restoration of approaches, following the strategy prerequisites to give data to the guarantors. Claims cover serving notification of a claim, keeping accounts, accepting and advancing any repair work, acquiring payment, and so on. (Alistair Barros, n.d.) The subject is an authority region and it is normal to have the proprietor's and contractual worker's specialists set up this site procedure.

Security procedures

Security procedures are frequently in light of the proprietor's current prerequisites which the contractual worker may create to suit his needs. Subjects, for example, the issue of passes, confirming of staff and the powers to look for are normal.

PEOPLE AND PROCEDURES

Every single great procedure ought to plainly characterize which individuals are required to do under the procedure. They regularly cover the accompanying:

- Giving of powers to a named individual or people;
- Offering powers to a man satisfying a specific occupation title;
- Characterizing the genres / trades of the individuals who can do the procedure; characterizing the different obligations of those specified in the procedure.

Structures are regularly arranged which demonstrate the detailing lines and obligations for a specific procedure. The structure should cover work title, name of people who complete the occupation, phone number and area. Frequently they incorporate the names of option individuals who are approved to complete the assignments. In specific conditions these structures may cover increasingly that one procedure when they cover a comparative theme.

It is normal to have similar individual complete different occupations inside various structures, yet mind must be taken to guarantee that there is no irreconcilable circumstance between the individual's parts. For instance, the individual who is in charge of accomplishing advancement of the work ought not to be in charge of health and safety.

Having the correct structure set up which is not overly complex and which incorporates suitably experienced and prepared individuals will bolster the smooth running of a project. Because of the generally here and now nature of projects and the evolving exercises completed in the term of a project it is normal for the structures to require customary updates.

CHANGES TO PROCEDURES

It is normal for procedures to must be changed amid a project. The progressions differ from minor ones which can be foreordained (e.g. a change in the contractual worker's site construction group because of the arranged retirement of a person) to major and unforeseeable changes (e.g. the local governing bodies all of a sudden forcing both weight limitations on vehicles utilizing certain locals get to streets and the hours that substantial great vehicles can utilize the streets). These progressions ought to be overseen accurately as per the built up QA procedures.

At the point when a change is being thought of it as must be told right off first as a potential change and its money related and time suggestions considered and controlled.

COMPLIANCE

Non-consistence of procedures can happen even on the best run projects. On the revelation of the problem it will frequently be important to advise any deformity and strengthen the procedure including giving extra preparing, and so forth.

The consequences of non-compliance problem can include:

- Expulsion of staff (for smoking);
- Stopping of the work on the basis of health and safety
- The contractual worker being expelled from future delicate records because of inability to effectively execute site procedures;
- Lessened payments because of not providing notification to the right time frame

CONCLUSION

Procedures are fundamental for various reason yet they should be clear and important all together that they bolster the project and don't limit it with organization management.

INTRODUCTION TO VALUE MANAGEMENT

Value management (VM), value engineering (VE) and value examination (VA), are terms that are being used gradually as regularly as could be expected under the circumstances; however, what do they mean? Most design associations have a stress for value and have methods and plan surveys to address overseer fundamentals and the cost of any proposed scheme. Capital cost is positively one part of value however expansive experience reveals that the slightest expensive is not for the most part the best and a more extensive understanding of value is required. (Adedeji Badiru, n.d.)

It is gradually perceived that not exclusively do design associations need to give assets that meet the owner's specific and are attempting to time and secret spending design however that the potential project increases the advantage to the owner's business. At the latest periods of project thinking the owner assurance is an announcement of various perceived and for honest to goodness 'necessities'. As procedure proceeds and the cost of treatment to those prerequisites develop it is habitually essential to reuse the planning to give an arrangement that is more acceptable to the owner. So also, as design proceeds with, additional open entryways can be made that could convey additional advantages to the owner's business.

It is a key piece of the procedure Association to enter this level headed dialog of value and need at the latest periods of the design. Experience exhibits that 'free discourse' and 'asking the manager what is required' is presumably not going to be acceptable, and design associations ought to have the ability to sink toward to the owner, measure the necessities and give plans which give incredible capital value and additionally improve the business.

It is this presentation of need and cost in another casing that VM strategies address. VM and related strategies should not to be seen as evacuating run of the mill design reviews but instead as giving a substitute perspective to the owner which is not evident from the common engineering assurance and cost evaluate. (The Joint Development Board sponsored by the Association of Cost Engineers and the Royal Institute of Chartered Surveyors, n.d.)

DEFINITION OF VALUE

Value can be showed as the connection between satisfaction of need and the cost of resources required to achieve that satisfaction. This can be connected as:

$$\text{Value} = \frac{\text{Satisfaction of need}}{\text{Cost of Resources}}$$

By enhancing the level of satisfaction and additionally decreasing the resources needed. At times, the level of satisfaction may be improved to such an extent, to the point that an expansion in resources is supported. Much of the time, the capital cost is an essential asset however working, lifecycle costs and availability of abilities could figure in the condition. It merits perceiving that diverse interests in a venture have individual needs; for example, a processing plant director could be searching for simplicity of support while the business supervisor may value a snappier request lead time. VM methods try to give systems by which these varying interests and needs can be united.

VM DEFINITIONS

Despite the fact that the terms value management, value investigation, and value engineering are broadly utilized; there is not add up to a concern to definitions. The European Union is chipping away at an overall way to deal with the understanding, practice and preparing for value management inside the EU and clarification of terms are normal.

However, it is by and large acknowledged that VM covers a scope of value enhancing grooms, especially those used for early reasonable considering and the “softer” front end issues of a venture, while VA and VE are all the more generally used to cover investigations of the conventional “more difficult” issues at the later phases of outline. Regularly, the term VE is utilized for learns at the venture cost level with VA used to portray more definite segment examination. (The Joint

Development Board sponsored by the Association of Cost Engineers and the Royal Institute of Chartered Surveyors, n.d.)

However, a few associations transpose these terms. It is conceivable to consider four wide levels of audit as appeared in Figure below.

The cost of progress at the later phases of configuration is likewise more prominent as procedure continues. Albeit self-evident, this point is often overlooked and desires can be made that VENA learns at the later phases of configuration will deliver considerable cost-advantage openings. This is improbable as real changes can't, as a rule, be endured on the grounds that the postponement and cost of rolling out improvements are unsuitable. This often brings about just minor changes of a cost cutting nature being made, which ought to have been recognized by ordinary outline/cost audits. In this manner, the distinctive commitment that VM can make is not understood, and at the very least the VM procedure is devalued.

The audits can be portrayed as beneath:

- An innovation/business audit confirms the business procedure and sets up a wide understanding of the innovation and overall working and different necessities.
- A theoretical survey sets up or affirms the overall venture scope and, if requests of cost are available, affirms that the wide plan and prerequisites are viewed as 'great value'.
- A extend cost survey is a more natty gritty examination of a proposed conspire completed when the particular engineering components have been determined and can be taken a gander at to affirm that the outline is ideal in
- A plant thing or part survey considers the outline of particular things.

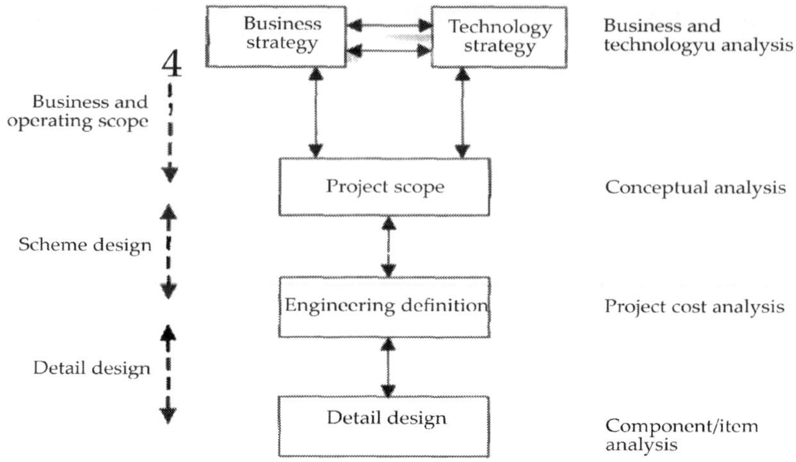

Figure 8 1 Review levels explained (The Joint Development Board sponsored by the Association of Cost Engineers and the Royal Institute of Chartered Surveyors, n.d.)

TIMING OF STUDIES

The chance to roll out reserve funds or different improvements decreases as the venture approaches detail outline and money related approval is given for buy of hardware and other engineering works - see Figure below.

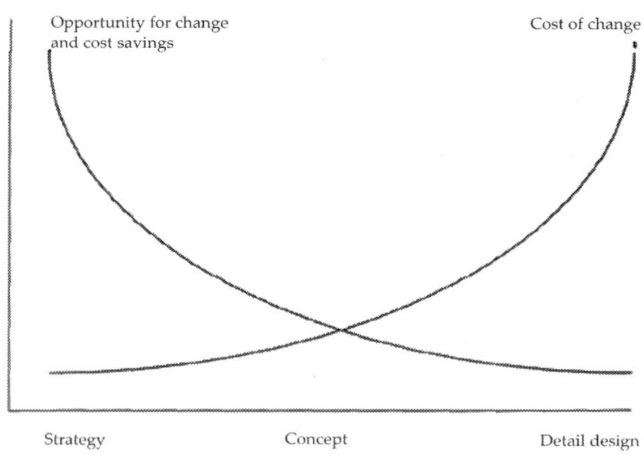

Figure 8 2: oppurtunity and change cost dependence in projects (The Joint Development Board sponsored by the Association of Cost Engineers and the Royal Institute of Chartered Surveyors, n.d.)

VALUE MANAGEMENT METHODOLOGY AND JOB PLAN

Commonly, VM grasps three components which together give the uniqueness of the approach. These are:

1. Team based working;
2. A organized 'employment design';
3. Use of the standard of "functionality" or 'need'.

The initial two are not one of a kind and are qualities of numerous critical thinking techniques. The last is the recognizing highlight of most types of VMIVENA techniques. "Functionality" is utilized to portray what the part, system, plant or venture is endeavoring to achieve. For instance, the "capacity" of an entryway may be to give privacy, give security, and make the correct picture or a combination of each of the three. This qualification gave by utilizing "functional" representations are not obvious from the engineering depiction 'entryway'. Representing sections along these lines empowers costs of things satisfying a similar capacity to be included, subsequently enabling the owner to understand the overall cost of that capacity. Taking the past design, it may be that if a component of the entrance is fundamentally for a picture then different components that are there for similar reasons can similarly be incorporated inside the functional portrayal. A superior understanding of the overall costs and alternatives for giving the picture capacity can be evaluated - including that for the dealing with executive's office! "Functional" depictions additionally vary from conventional engineering portrayals in that numerous things of hardware satisfy more than one capacity without this qualification being clear from the cost evaluate. (Bessant, n.d.)

The rule of functionality can similarly be utilized to help develop the extent of a proposed extend. An effective concoction plant venture won't simply have the fundamental procedure capacity of 'respond feedstock's' however will likely have different process capacities, for example, 'recover materials', 'remove pollutions', 'achieve item shape', and so on. There will likewise be capacities, for example, 'guarantee operability', 'make safe', 'ensure nature', 'accommodate the future', 'improve client benefit', and so on. These are for the most part run of the mill capacities for a procedure plant, the costs of which are not clear from a typical engineering detail and cost gauge. It is these different capacities that decide if a benefit will be 'world class' or in front of the opposition and not only a duplicate of that being utilized by others. It is basic, in this manner, that these capacities are appropriately characterized and investigated. It is typical to show functional depictions as a FAST - functional analysis system technique - diagram. A component of this kind of diagram is that the progressive system of needs/capacities empowers costs and chances to be comprehended at various levels. A case of a FAST diagram

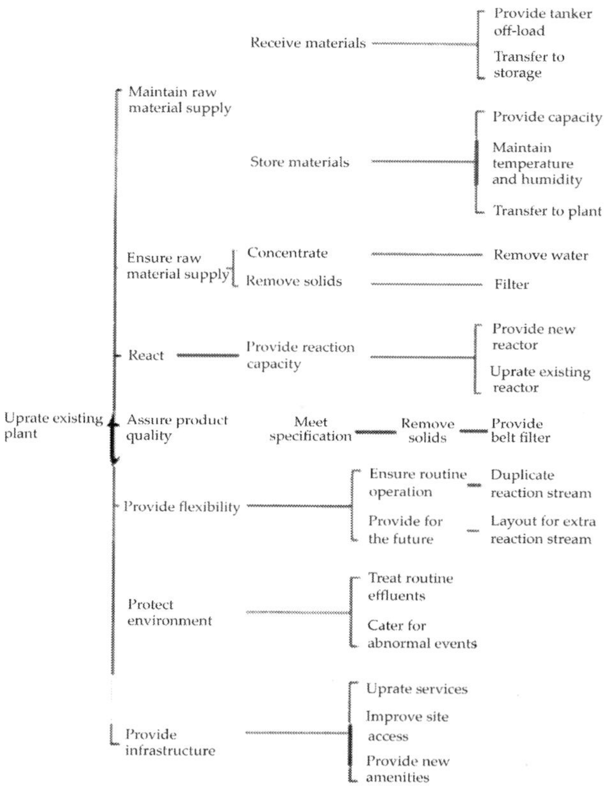

Figure 8 3: A FAST Diagram (The Joint Development Board sponsored by the Association of Cost Engineers and the Royal Institute of Chartered Surveyors, n.d.)

For a chemical plant is appeared in Figure above. From this it can be seen that 'guarantee item quality' could be considered at various levels, i.e. change the item determination, change the procedure innovation to give a more effective process, change the hardware sort or change the detail plan of the chose gear. These present open doors without breaking a sweat of usage and exchange off. This type of introduction gives an upgraded perspective of the venture that is not evident from a typical engineering gauge.

An element of the FAST diagram is that it is developed from the left-hand side by asking "how?" and from the right-hand side by asking "why?" It is this scrutinizing empowers the principal reason to be set up. For example, asking 'why a refining section?' could give a variety of answers including: 'remove polluting influences', 'recover material' or 'focus the item'. These diverse portrayals give a very alternate point of view on the need for a refining section and the conceivable alternatives, e.g. diverse process, gear, working (get another person to do it) or business. A further element of the FAST diagram which prompts more prominent lucidity is the utilization of 'verb-thing' depictions of capacities. For example 'increment item quality', may most likely require a very extraordinary arrangement from 'keep up item quality'.

The job plan

The job plan works in a requested and systematic way. The primary phases of the job plan that are generally utilized amid outline development are as per the following.

Information phase

The information phase involves characterizing the project, gathering the available information, setting up requirements and completing a capacity analysis to depict 'What is it we need the venture to do?' rather than 'What are the physical segments of the venture?' It is toward begin of this phase the level of study and potential team individuals ought to be characterized.

Creative phase

Amid the creative phase the team utilizes the functional portrayals, regularly displayed as a "capacity" or 'FAST diagram' to investigate alternative courses for finishing the coveted capacities. "Conceptualizing" is normally used to produce thoughts.

Evaluation and analytical phase

The thoughts created amid the past phase are screened and evaluated by the team amid the evaluation and analytical phase. Ordinarily, some type of choice evaluation technique would be utilized to help amid this phase.

Development and recommendation phase

Amid the development and recommendation phase the team surveys the chose thoughts and concurs on the proposition to be received or requiring further investigation.

Introduction phase

The introduction phase involves the introduction of the concurred outline proposition including any outstanding developments.

Implementation phase

See that the recommendations are developed and advanced amid the implementation phase. Often, a considerable measure of vitality is put into the initial five phases over a genuinely brief period however as implementation is over a more extended period and involves individuals not involved in the underlying survey, energy can be lost.

The job plan is not intended to be prescriptive but rather is there to enable a team to work systematically through the outline to phase. Judgment is needed to structure the job plan to suit the level of study and multifaceted nature of the venture.

The study team and resources

There is a view that reviews interfere with the outline procedure. The inverse is the situation. Appropriately set up considers empower the invested individuals to meet up and direct a comprehensive and organized audit of the venture, especially at the beginning periods of outline. By and large the study can replace typical outline audits and team gatherings. Surveys exhibit the open door for individuals not ordinarily occupied with the outline procedure to be involved. It is just the owner and invested individuals that speak to the owner's various advantages that can change the extent of the venture, the engineering team can make recommendations, and then just on the off chance that they understand the more extensive parts of the venture.

Concerning most team based exercises, the extent of the study is vital. An excessive number of or excessively couple of members brings about the study being unmanageable or there not being adequate cooperation of thoughts. A team of five to eight individuals is perfect and choosing the team often gives a decent understanding into who is truly essential in the venture choice process.

The formal team based phases of a study, i.e. development of the capacity diagram, hypothesis and evaluation, can typically be completed over a half to two days for each phase with breaks to compress and produce more information. Subsequently, over a time of one to two weeks, most little to medium sized tasks can be checked on. Now and again a very definite analysis of an extensive plan down to the segment level will take longer, however note needs to be made of past remarks on the chance to roll out noteworthy improvements. (George A. Steiner, n.d.)

CONCLUSION

In spite of the fact that the value management techniques that have been portrayed depend on entrenched and proven techniques there is the need to perceive that the experience and reputation of the 'study pioneer' is basic to progress. For any person to wind up noticeably proficient and reasonably experienced it is suggested that study pioneers ought to be completing around one study a month. There is the allurement to believe that architects and originators who have gone to a VM instructional class can lead VM surveys. The experience needed to guarantee a sound theoretical understanding of the techniques can just - as in any movement - be increased through training and the individual discovering that happens. The techniques need to be utilized adaptably and with creative energy: an excessively unthinking application can here and there miss the issue and the study pioneer needs to know when to adjust the technique to draw out the main problems. This requires the study pioneer to have the capacity to draw on other techniques and procedures to help the development of answers for the issues raised by the formal "capacity" analysis.

CHAPTER 9

Operational Maintenance

INTRODUCTION

Capital and revenue expenditure

At the point when a plant is fabricated, the costs for the design and construction are usually made available from a capital budget for investment in new assets. One feature of the design phase is the consideration that must be given to future running and maintenance aspects. The owner's loss in production, and therefore earnings, during a shutdown, can be enormous, and maximization of reliability and reduction of maintenance effort must be incorporated into the plant design. This is a subject in itself and is not secured here.

The plant will be worked by a construction team, usually through a contractual worker, who will then hand over the plant to the commissioning team. Commissioning means starting the plant up and running it under normal operating conditions to watch that the construction and the process conform to the specification and design. Commissioning work is exceptionally specialized and complex and won't be dealt with in this record. The costs of the commissioning phase form part of the capital budget.

Once the plant is producing and meeting the process specification in quantity and quality, the plant is handed over to the organization which will operate and maintain it on a day-to-day basis. This organization is normally the owner, however may be a separate company which operates the plant without owning it. Starting here onwards the costs are not capital costs, but rather are "operating" or "revenue" costs.

A budget is normally set aside each year to take care of the running costs of the plant, part of which is the operating expenses, i.e. salaries for the operators, power, oil, gas and water to keep the plant running, and the feedstock which is the raw materials from which the item is made. The other part of the operating budget, and the part with which this chapter is mostly concerned, is the maintenance budget. This is the money required for repairs to, and servicing of, the plant.

Maintenance can be further split into two subsections; the first is the place the repairs and servicing can be carried out while the plant is still running and producing; the second section is the place the plant must be stopped and made safe before the repairs and servicing can be performed.

There are many names for each sort of maintenance yet we will distinguish them here by calling the first 'routine maintenance' and the second 'shutdowns.

In the past, the generally accepted principle for maintenance, either routine or shutdowns, was to carry out preventive maintenance, the theory being that if the equipment was kept in the best working order then the incidence of breakdown would be decreased. The present thinking, which is reported to have started in the USA, is that some aspects of preventive maintenance make it an exceptionally expensive and unreliable technique for risk prevention, and that it is more cost successful to repair equipment only when a breakdown occurs. ('On the off chance that it isn't penniless, don't settle it,' is the US catch phrase for this concept.)

Routine maintenance

This is sometimes alluded to as 'running maintenance', 'on the run maintenance' and 'day-to-day maintenance' among many other names.

Routine maintenance covers all aspects of work required to ensure that the plant runs smoothly and effectively and ranges from grass cutting of landscaped areas to window cleaning, from light changing to major electrical works and from vehicle maintenance to complex compressor/ turbine alignment. There are various reasons why maintenance needs to be carried out, and these will be shrouded in greater detail later in this chapter, however in essence may include the following:

- Items may cease to work
- Items may work inefficiently
- Items may work unsafely
- Items may work unreliably
- Items may be costing too much to operate due to a fault in design or performance; and
- Items may be creating unacceptable environmental conditions such as odor, smoke, greenhouse gasses, pollutant exhaust gasses, and so forth.

The decision as to whether maintenance work can be camped out while the plant is running would normally have taken place when the plant was first designed and would rely on such factors as the cost of duplication, the criticality of the thing and the kind of process, e.g. continuous running or batch processing. Examples of continuous running would be refineries or coal-fired power stations where starting and stopping are exceptionally expensive.

Batch processing is every now and again found within the nourishment industry (e.g. bread baking), in the pharmaceutical industry or iron foundries. With batch processing, maintenance can generally be performed between batches.

Normal maintenance must still be carried out, in any case, to ensure that the equipment is functional when required.

In the case of continuous processes, there is no opportunity to carry out maintenance between batches and therefore the design of the plant must consider all practical options to decrease the risk of one thing disrupting the whole production.

In a refinery, for example, it may be cost-viable to construct a section with two pumps where only one is actually required and one is spare.

These pumps are usually alluded to as "obligation" and "standby" respectively. This option allows one pump to be maintained while the other one works.

In spite of the fact that this might be pragmatic with certain little things, it would be restrictively costly to outline a plant with an extra high weight, medium weight and low-weight turbine/compressor setup in case one may separate amid operation. Consequently, hazard and cost are a central point in outline criteria with respect to maintenance contemplations.

Institutionalization of hardware is another fundamental, empowering a owner to keep just a single substitution for each kind of unit as opposed to one for every thing of the plant, affecting reserve funds in income and capacity costs.

Shutdowns

Again there are a few option names given for shutdowns. Basic names incorporate shut down, blackout, update, pivot, stop or different appellations. For this situation, we will consider all to mean the same, albeit a few associations and to be sure people have distinctive implications for each.

Some view a shutdown as just when the plant is ceased at long last prepared for disassembling, while others view a shutdown as when the plant stops out of the blue. It again depends, particularly on the person.

Arranged shutdowns are essential where maintenance work is impossible as routine maintenance while the plant is running. Cases of this are heaters or boilers in control stations and refining or fractionation towers in refineries. As a rule, shutdowns are not required by the owner or even as a prerequisite of the procedure yet are real to acclimate with the statute. There are distinctive laws administering different divisions of industry, however, for the most part, these laws have advanced to guarantee that the owners, as well as the general population, are protected from the disappointment of any hardware and ensuring spillages or blasts. One of the primary operations completed amid shutdowns is, accordingly, obligatory examination to agree to these laws.

Many procedures cause fouling of hardware and this must be cleaned paying little respect to lawful prerequisites. Another regular purpose behind shutdowns is to consolidate upgrades in the process or debottlenecking of the plant. As the quality of a chain is reliant on its weakest connection, so the throughput of a plant is administered by the most compelled thing of gear. This is known as the bottleneck. Debottlenecking is the expulsion or diminishment of these requirements (this can't be rehashed forever as the plant is just intended to a specific limit and ceaseless redesigning turns out to be less savvy). (Tonchia, 2008)

The planning of a shutdown is typically represented by statute and market powers. The law may conceivably be extended by formally applying for a postponement, yet advertise powers may not. It is normal to design well ahead of time when the following shutdown will occur. All occupations which can't be finished with the plant running will at that point be recorded as conceivable employments for incorporation in the shutdown works. Work on the shutdown will be constrained quite far, to limit the loss of creation and benefit. The most limited shutdown time frame is the ideal target.

PLANNING

Scheduled routine maintenance

Scheduled routine maintenance can be identified with the consistent adjusting of an auto. It is a similar idea in any procedure plant; the segments might be unique, yet the standard continues as before. Pumps may require new seals, valves require new pressing or situates, or other consumable things may require supplanting. This work can regularly be completed either while the plant is running, or amid a plant shutdown. For each situation, it is indispensable that the maintenance is appropriately arranged.

With a specific end goal to do certain maintenance while the plant is running, there may must be a pre-introduced go down segment, for instance, an optional standby pump which can be run while the primary pump is destroyed. This work must be deliberately intended to guarantee that maintenance is done before the parts achieve such a state, to the point that the gear is harmed or bombs totally. The organizers must devise a calendar, ascertaining the ideal running time of every segment which requires to be adjusted, and circulating the undertakings to maintain a strategic distance from pinnacles and troughs in workload. (Bessant, n.d.)

The methods for planning are regularly checked to find out the measure of work in worker hours to help decide the base number of staff required to keep the plant practical. It is no utilization creating a calendar which requires 140 men each third seven day stretch of the year and just 15 men in the interceding weeks. This would be an astoundingly costly approach to utilize work, regardless of whether it is in house or provided by means of a contractual worker.

In case of a scheduled shutdown, much more highlighting is required in planning. Many plants either encourage, or are nourished by, different plants, and these different plants must be considered when choosing the date and term of a shutdown. To give an illustration, oil establishments in the North Sea are consistently delivering oil and gas. The oil might be always required for refining and is the sole feedstock for a refinery. The gas delivered in the meantime might be isolated and channeled to an ethylene plant which extricates the ethylene and sends it to a polyethylene plant which makes polythene dots, which are thus sent to producers to make gas and water pipes or even polythene sacks for grocery stores. On the off chance that it is chosen to close down the ethylene plant, it must be chosen what happens to the tremendous amounts of gas as yet being created from the North Sea, how to nourish the polyethylene plant and how to keep up generation of channels. An undeniable answer is have vast storerooms for each stage, yet this is an uneconomic arrangement given the low costs, the opposition for every item, and the massive stockpiling zones which would be required.

Organizers should in this manner liaise intimately with the "business" units of each plant. These units care for the owner's business advantages, for example, choosing from where the feedstock and go down provisions might be purchased, and what their promoting duties are for their own items. The entire chain must be counseled before a date can be orchestrated. Generally the specialty units manage to the organizers when every unit can be closed down.

The following thought is the length of the shutdown. Once more, this is often set by the specialty units. They might have the capacity to purchase feedstock at low costs for a specific period from elective sources, or they might be given a time of allotment amid which their clients will discover elective providers. Contract duties and market powers are generally immovably set with respect to these periods. It is thusly essential that all work on a shutdown, or the part of work which must be done amid a shutdown, is wanted to be completed inside the given time. (Horngren, n.d.)

The work required, for example, to vessel internals, can more often than not be surveyed in view of authentic data, yet a few adjustments to the program might be vital should the degree of work observed, on opening up, to be pretty much than foreseen.

To keep the shutdown time frame to a base each assignment must be explored and thought given to techniques for working which will augment the preparatory work, construction and so forth which can be attempted before the plant is closed down. Utilizing elective techniques for work which decrease the workload in the shutdown time frame may build the aggregate cost, and this must be adjusted against the diminished time that the plant is non-beneficial.

The organizers, thusly, need to take a gander at each undertaking to be completed and ascertain, utilizing man-hour standards, to what extent each assignment will take. They should survey what number of men can work at the same time on each errand, accordingly working out the term of those undertakings. Many plants are exceptionally perplexing and congested and it is not generally conceivable to utilize the hypothetical ideal keeping an eye on levels for reasons of safe working and openness. There are just such a large number of men that can be utilized proficiently on each assignment.

The organizers must take a gander at all the errands, the quantity of men who can deal with each, the span of each, and what number of undertakings can run at the same time.

Certain errands rely upon the past one being begun or finished before work can begin. The ideal will in the long run be settled by utilizing different planning strategies which it is not fitting to depict here; suffice it to state that planning is a science all alone. There will be one grouping of assignments which can't be decreased in term and is subject to different errands wrapping up. This is known as the basic way. The basic way can't be abbreviated or decreased and this decides the length of the genuine shutdown. The term given by the business gathering might be adequate to do the work under typical move working or may even require seven day working, 24 hours for every day. Where the basic way can't be fitted into the time scale, at that point transactions are required to sensibly broaden the time permitted. (Adedeji Badiru, n.d.)

With great planning and basic way examination consolidated with the time permitted by the specialty unit, the organizers in conjunction with expert designers can work out the aggregate number of men required on the site and the move design they are required to work. This data is basic to contractual workers for resourcing the errand, and to the cost control staff for settling spending plans.

Scheduled particular maintenance

Scheduled particular maintenance can be depicted as a noteworthy component of work which was not visualized in the long haul running of a plant. The work is basic and can be scheduled to be done close by the routine maintenance while the plant is as yet running. Particular maintenance dependably puts a strain on assets, and powerful planning of the work to fit in with other maintenance is basic. A case of particular maintenance would be the place dampness has entered 'between protection materials and pipework. Because of fine activity, this water will spread over a huge number of meters, making erosion the pipework, which will be subtle to the stripped eye as the external protection will be unaffected and the purpose of entrance might be hidden (under lagging consumption). ,

The maintenance required here will be the stripping of all protection, deciding if the pipework can be impacted and repainted or must be supplanted, and the re-insulation of the funneling (perhaps to a higher determination) to forestall repeat. The stripping and re-insulation could be performed while the plant is as yet running, given the protection are not basic to the procedure.

The planning of this particular maintenance must consider the protection contractual workers' accessible assets, the contractual workers' normal workload, regardless of whether a discrete contract ought to be granted and work by different contractual workers in a similar region. Should an underestimation be made with respect to other protection work being simultaneously attempted by the contractual worker, at that point overwhelms of expenses and time, and even loss of creation, may happen?

Exceptional maintenance

It is frequently hard to recognize exceptional maintenance and particular maintenance and numerous owners utilize just a single articulation to cover the two classes. The cover regularly makes the refinement incomprehensible notwithstanding when the two portrayals are utilized. Exceptional maintenance might be depicted as maintenance which was not expected in the long haul running of the plant (the same as particular maintenance), however, is not scheduled and can't be done over a long stretch. Exceptional maintenance might be done exclusively to expect breakdowns. A representation of breakdown maintenance is given in the accompanying area. (Bessant, n.d.)

A case of exceptional maintenance would be the place the feedstock of a plant change, making a pump end up noticeably stopped up with slop as the new conditions are outside the pump's outline criteria. The pump would turn out to be extremely wasteful and may influence the yield of the plant.

Planning the cleaning or substitution of this pump must not influence the routine maintenance of whatever is left of the plant. Once more, the accessibility and abilities of the contractual worker's assets must be considered, and furthermore, the simultaneous maintenance prerequisites must be anticipated.

Breakdown maintenance

As the name proposes, this is the maintenance required when a thing of hardware fizzles. On the off chance that the hardware has a standby, at that point generation won't be influenced and the repair time is not so basic; but rather if there is no standby the plant will stop creation. The planning of this work not just includes the repair and the conceivable restart of the plant, yet additionally the organizing and rescheduling of all other work which is conceded because of the basic idea of the breakdown.

Constraints

Process plants are by and large extremely congested with pipework, cabling, and hardware. The productivity of which work can be completed is particularly lessened in correlation with the development of another plant. Also, on a few plants, the nature of the chemicals and gasses contained in the process are amazingly perilous to well-being, and introduction to them is allowed for a constrained period even with defensive attire and breathing contraption. (Horngren, n.d.)

The idea of work on live plants, for example, working inside vessels and towers which are exceptionally confined, and the prerequisite to have someone else remaining by outside for security reasons, and for working in defensive suits with breathing contraption, are on the whole segments of extremely wasteful working. Different constraints incorporate departure of the region preceding radiography of welds, confinements on fire cutting of pipework or steel, the utilization of non-starting apparatuses and welding just in named safe zones, all of which must be assessed when planning the work. (Waldemar Karwowski, n.d.)

The above circumstances, to differing degrees, are on the whole present in process plants and it is the organizer's business to compute how every circumstance influences the man-hour standard and efficiency. Each plant or part of a plant will have an alternate factor to apply to standards to build up a precise general arrangement. A precise gauge of the cost is to a great degree troublesome.

Coordination

Similarly, as with another venture, a program must be created for all the known maintenance work. Should a breakdown happen, or exceptional maintenance be required, this program may be incidentally surrendered for the more basic work to be performed. This frequently requires the contribution of more than one contractual worker. For one mechanical occupation to be executed there is as often as possible a prerequisite for scaffolds, covers, electrical contractual workers, monitors, printers, and non-ruinous analyzers. The operation of these contractual workers must be entirely planned to limit the time taken by each furthermore, the obstruction with others. Notwithstanding when a vocation is basic, the time was taken by the organizer to deliver a program is more than spared by the expanded and effectiveness of the organized working. The arrangement of materials, spares, and parts all must be arranged, pre-requested and facilitated to guarantee the work is not postponed.

All agreements for maintenance work should address the measure of coordination required amongst contractual workers and with the owners and should make remittance for unexpected occurrences which require adaptable working. (Salvendy, n.d.)

ESTIMATING AND BUDGETING

Estimating, for the most part, is secured somewhere else, and this section points just to allude to estimating and budgeting particularly to maintenance.

Estimate

The estimate is a forecast of the last cost of work. In its planning, it requires a mix of involvement, historical data and a conjecture of business conditions connected to a specialized understanding of the work.

The exactness of the estimate will rely on the level of specialized definition, the nature of the estimating data and the evaluation of the dangers innate in the maintenance work including sensible recompenses for questions.

Sources of cost information

Historical data

Records of past maintenance work and buys more often than not contain broad and valuable cost information. On the off chance that such information is deliberately examined, recorded and documented together inside an appropriate recovery framework, this structures a decent reason for costing future work.

While using historical data as a premise of an estimate, it is essential to consider the accompanying focuses, both in connection with the present venture and to the venture from which the historical data were separated:

- Procurement system;

- Reliability of information;
- Productivity factor;
- Similarity of the arranged work (neighborhood conditions and time scale);
- Local conditions;
- Time scale;
- Condition of the plant.

Unit costs/norms databases

Estimating data for operational maintenance has by and large been troublesome, with every owner utilizing contrasting techniques for estimation, types of agreement and details. Not at all like the building business, there is minimal distributed information accessible as to value. In any case, distributed work norms exist for the lion's share of orders in the development of process plants, and these might be adjusted for use for maintenance work. (The Joint Development Board sponsored by the Association of Cost Engineers and the Royal Institute of Chartered Surveyors, n.d.)

The real distributions are created by:

Page and Nation;

- Oil and Chemical Plant Constructors Association (OCPCA).

When utilizing such information it is vital to consider all things included under 'historical data' and the accompanying:

- The costs/norms ought to be for maintenance work and not new development;
- The measure of work, e.g. a little amount of work will be all the more "costly" per unit than an extensive amount of work.

The above norms are by and large factorized as per the workplace and work efficiency.

Management costs

On new ventures, the management is as often as possible devoted to the work and accordingly all costs are charged to the venture. With maintenance, there may not be any staff committed to one particular zone or cost focus. While estimating the general costs of maintenance it is important to incorporate the costs of management in the right cost focuses, or their genuine costs may because cost overwhelms when at last apportioned.

Owner doesn't generally allow their own directors/staff costs to maintain, as these are perceived as an overhead cost to the business. For the situation where owner's staff are committed to the shutdown, other faculties are often contracted in to fill their typical parts. The costs of these contracted staff must be incorporated into the general shut down estimate in spite of the fact that they don't have all the earmarks of being an immediate cost. Correspondingly, and all the more clearly, if an overseeing contractual worker is named, the costs must be designated to the occupation being overseen.

Numbers of contractual workers

The quantity of contractual workers utilized on routine maintenance, or a shutdown, directly affects the costs thereof. For every contractual worker, there is a level of management inside the costs. In the event that few contracts can be included in one expensive contract, there is just a solitary management segment to be incorporated, though bigger than every individual one, however littler than the total. On the other hand, if the agreement turns out to be too expensive, the required contractual worker's management resources may require the utilization of a substantial multi-utilitarian contractual worker with the chaperone extensive overheads and this could be wrong for the sorts and measure of work required.

The utilization of numerous contractual workers includes more prominent coordination by the owner to avert interruption.

These contemplations must be considered while planning maintenance estimates, spending plans, and contract logic.

Reliability of costs

Because of the different factors and working conditions influencing maintenance, any historical costs or norms must be painstakingly assessed to guarantee that they are suitable. Basically taking a standard for supplanting seals in a pump would be mistaken if the man-hour standard was for a draw in greasing up oil benefit and a similar worker hours were being connected to take a shot at a direct in unfiltered unrefined petroleum benefit. Moreover, all costs must be checked to decide the development of that cost. (The Joint Development Board sponsored by the Association of Cost Engineers and the Royal Institute of Chartered Surveyors, n.d.)

Management costs may have been barred from singular gear costs on past employments yet are to be incorporated per thing of hardware in the estimate. It is likewise imperative to guarantee that the costs and costs in any historical data are real costs used and not just figure costs which were really invaded. Costs of spares and supplanting ought to be checked with the maker to guarantee that expensive value changes have not occurred since the past consumption. The presence of spares value concurrences with makers ought to likewise be researched.

Category of maintenance

For different reasons, numerous owners wish to know in what category of maintenance costs are being used. Estimates should be isolated into each of these classifications per thing of work. This arrangement is notwithstanding the cost focuses.

A few of the more typical classifications include:

- Reliability: maintenance is required in light of the fact that the thing of gear is temperamental in operation;
- Statutory: gear must be opened up and reviewed as per statutory prerequisites, e.g. Weight vessels;
- improvement: the gear expects change to enhance the item or process; investigation: examination division have asked for that the hardware is opened to check consumption, disintegration, and so on., for the most part, because of issues experienced in a comparative situation somewhere else;
- Safety: The Equipment Requires Inspection or Modification If There Is A plausibility of break, spillage, and so forth;

- Execution: the gear requires repair or change to enhance its execution or yield;
- Environment: the hardware requires some work to diminish contaminations being acquainted with the earth.

A portion of the above will make the plant more proficient and will pay for themselves in time, while others will have no money related advantage and should be composed off as an immediate misfortune, however, will have circuitous advantages, e.g. security, natural, and so forth.

CONTRACTING STRATEGIES

Introduction

One of the phases in drawing in a contractual worker to attempt maintenance work is the planning of the delicate reports. It is at this phase the clearly direct matter of picking the sort of agreement can impact the monetary accomplishment of the venture. An agreement system must be produced which will mirror the sort and measure of work, the many-sided quality, and reliability of the plant and the time accessible for acquisition and execution of the work.

Facilities management

Facilities management is a technique for maintenance where all the duty regarding the required accessibility of the plant is outsourced to a contractual worker.

Before offering the contractual worker is given the detail of the plant, any accessible historical data of the past maintenance and the working prerequisites. The contractual worker at that point presents a cost for keeping up the plant to a specific standard for a specific period. Installment terms might be singular amount or asset based. This sort of agreement is seldom utilized on bigger process plants, however, is winding up noticeably more typical on assembling and generation line plants. (Bessant, n.d.)

Routine maintenance

By it's extremely nature, routine maintenance work will cover arranged and impromptu exercises, work which is both quantifiable and unquantifiable and works which may be attempted outside typical working hours.

In all cases, the central prerequisites of any agreement ought to be 'adaptability'. Because of the eccentrics of breakdowns, even with the best planning, pinnacles and troughs in workloads are sure to happen and any contracting technique should address and take into account these varieties

Shutdowns

To anticipate the pointless loss of generation it is basic that shutdown time is kept to a minimum with however much preliminary work as could reasonably be expected being performed before the plant stoppage. Any shutdown contract ought to be adaptable and require the contractual worker to supply extra resources for any additional work arising as vessels are opened and inspected. The contractual worker would even now be required to finish the work without an augmentation of time.

Shutdowns are either those planned for maintenance or the unplanned stoppage of the plant resulting from a breakdown.

Planned/unplanned

Shutdowns resulting from planned exercises can be suited for a type of agreement where control can be connected through a calendar of rates and for reimbursable costs. Single amount contracts are seldom effective because of the likelihood of rising work on opening gear. Unplanned shutdowns may require an absolutely adaptable approach where a reimbursable contract might be the main choice.

Term contractual workers

Existing term contractual workers can be utilized for work on shutdowns. In any case, there might be a peril of duplication of resources, i.e. payment by worker hours on one contract while the same is being paid on a timetable of rates somewhere else. Furthermore, term contractual workers might not have the resources or abilities to do shutdown work.

Partnering

Reference ought to be made to partnering as depicted somewhere else in this book. This approach is not prescribed for shutdowns because of their relative here and now nature and the long haul pre requisites of partnering unless the shutdown shapes some portion of a bigger maintenance contract.

In-house

An in-house maintenance office will once in a while have the ability to handle the entire shutdown in the time required or it will be executed to the detriment of other maintenance. It is more regular that in-house maintenance handles discrete regions of the shutdown where authority information is required, e.g. instrumentation, turbines/compressors.

Management contractual worker

The administrations of a management contractual worker are often incorporated into the management of the shutdown. The appointment ought to be as ahead of schedule as conceivable to empower legitimate planning and coordination of the work. The management contractual worker gives the main management group and can likewise be the executing contractual worker at times.

Methods of payment

For the most part, the payment for maintenance or shutdown contracts can be categorized as one of the following three main classes:

- Reimbursable (essential cost in addition to management charge);
- Lump total;
- Schedule of rates.

There are numerous subordinates of these payment methods, and contracts can contain more than one category. Reference ought to be made to Chapter 7 on contract conditions for more detail of these agreement sorts.

Disaster recovery

This is the approach and the exercises required to bring a shutdown or maintenance work, either back on the program or within spending plan or both. The principal necessities will be to distinguish the problem(s), organize the timetable, and dispense resources, set up a definitive reporting and monitoring framework and continually audit advance against basic exercises.

Each agreement put for shutdowns or routine maintenance ought to contain arrangements to take into account this restorative activity without determining the agreement.

CONTROL OF COSTS

Introduction

The basics of cost control on maintenance ventures are:

- Cost monitoring: a methodology fit for providing dependable information in due time;
- Cost control: the basic leadership in light of the information gave by the cost monitoring system.

Cost information is required for maintenance extend consistently:

- Daily and/or
- Weekly and/or
- Monthly.

Without this information, management has lost a noteworthy device.

The main segments of cost control are:

- Good estimating method;
- Effective cost code or cost focus framework;
- Workable variety framework;
- Efficient 'genuine cost' reporting framework;
- Practical and powerful methodology for predicting patterns and final costs.

It is the last segment that uncovers where things are going incorrectly. Without this segment the framework only screens. To get to this fifth stage one must advance through the initial four phases and then apply judgments.

Where essential, choices and changes are made to rectify any issues to bring the occupation back on to time and spending plan.

Work orders

On term maintenance contracts and sporadically on shut down contracts there is no definitive extent of work contained within the agreement archives. The contractual worker is required to perform work as and when instructed by the engineer.

To control costs on these sorts of the agreement, there must be a composed instruction to the contractual worker which unmistakably portrays the work, gives the starting date and time, and distinguishes the strategy for repayment and to where the costs must be apportioned. This information is essential for budgetary control on particular things of hardware, forms or topographical regions of the plant, depending on the cost reporting prerequisites. For historical purposes, it is additionally fundamental that the correct work scope, costs and points of interest are recorded, with the goal that future estimates will be precise. (George A. Steiner, n.d.)

The most effective technique for instructing the contractual worker is by a work arrange framework. There are a few names utilized for a work arrange, for example, site instruction, contractual worker's instruction, specialist to perform work, and so forth.

The obligation regarding the raising and issuing of work orders is that of the engineer/supe~isor in charge of the work.

Variations

The control estimate ought to contain a contingency total which will include for some rising work which is probably going to be distinguished on opening and inspecting the hardware. In the maintenance area, in any case, unexpected things of work and crisis repairs regularly happen. As this work is not within the anticipated degree there might be no arrangement for it in the financial plan.

The venture chief must choose if the cost of these extension changes can be contained in the contingency aggregate. Assuming this is the case, stores must be exchanged out of the contingency to the pertinent cost focus.

On the off chance that the contingency entirety is insufficient to suit the degree changes, the venture owner should either apply for a supplementary endorse or demonstrate a cost invade. It is basic for a change technique to be actualized with a specific end goal to run down and record any varieties which may happen. To control the maintenance costs, all progressions are recorded and their belongings legitimately assessed.

The program and spending plan would then be able to be refreshed as the progressions are approved. Contractual workers actually construct extraordinary significance in light of recognizing changes which will influence cost and program and will cause interface issues with different disciplines.

Varieties can often bring about the interruption of the contractual worker's work with ensuing potential for claims for extra costs and deferrals.

Measurement

Measurement is important to determine the genuine work scope against which costs are used in a wide range of agreement aside from single amount. There is a school of thought, in any case, which advocates measurement of resources utilized and work performed on singular amount work to determine an incentive for cash and to aid the precision of future estimates. However, contractual workers are hesitant to use exertion in preparing these figures. Where these figures are required, it ought to be expressed as a commitment of the contractual worker in the delicate records.

Measurement of maintenance work, for the most part, should be done at the earliest opportunity as, not at all like new development extends, the work is once in a while appeared on drawings and huge numbers of the materials are reused as opposed to new. It along these lines winds up plainly hard to distinguish work everything considered. Correspondingly, if work is being done on a reimbursable premise, it is important to screen and report resources per thing, and this is impossible by and large if records are incomplete. A portion of the mechanical or specific work might be performed inside vessels or towers, and it is typically useful to shut everything down the gear when the work is finished, which makes consequent measurement unimaginable.

Unless measurement is precisely performed and recorded the reliability of cost information is disabled. This, in turn, lessens the precision of the cost reports and future estimates.

Approval of work

Payment of anything of work is based on the work being acknowledged by the owner. During maintenance work, the acknowledgment can sometimes turn into a hazy area. In the event that a contractual worker bids a singular amount for opening a vessel and cleaning for inspection, the level of cleanliness must be specified or a disagreement could arise. The contractual worker will wish extra payment in the event that he is asked to clean further, while inspection personnel will feel that the cleanliness they require is secured within the original single amount.

On reimbursable contracts, consideration must be paid to rework in instances where the original work is rejected. In the event that a contractual worker fits bolts of the wrong material and then has to refit the right bolts, the ideal opportunity for removing and refitting should not be payable, unless because of the owner's mistake.

Star rates versus day works

Contractual workers may delicate exceptionally aggressive rates in the schedule of rates keeping in mind the end goal to win work on a long haul maintenance contract. To minimize this risk, the contractual worker may then request the use of reimbursable rates on variations instead of the unit rates. Conversely, the owner may have neglected to define enough some elements of the work, yet will nevertheless attempt to apply the unit rates with a specific end goal to minimize his own particular risk.

Where an estimate is based on prevailing unit rates, cost control could suffer enormously should most of the work be paid on a reimbursable basis? For this reason, it is essential that a great deal of effort is used in the composition of the schedule of rates to ensure that the items are comprehensive and precisely describe the kind of work to be performed.

Disruption of contractual worker's work and potential for claims

Because of the variable character of maintenance, uninterrupted work is not always possible if different breakdowns happen, either in the vicinity of the work, or requiring the use of the contractual worker's resources elsewhere.

Disruption on a live plant could be caused by the presence of hazardous substances, lethal gasses or other safety restraints. Every one of these interruptions means additional costs to the contractual worker which he will pass on to the owner, either in the form of additional work or as a claim for additional costs and delays. (Alistair Barros, n.d.)Where the delays surpass the expectations included in the estimate, there could be a necessity for contract variations to be raised, albeit no additional work or changes to scope may have happened. One of the difficulties in precisely forecasting final costs is assessing the probability of contractual workers submitting claims for such delays and disruptions toward the finish of the employment, and the settlement estimation of these claims. Owner's representatives should precisely screen all delays and disruptions to help assess and concur any such subsequent claims for extra costs.

Self-policing by contractual worker

Where practicable, some responsibility should be set on the contractual worker for assisting in the cost control of the maintenance. Obligations and incentives should be included in the contract report to this impact. The greatest incentive for this control is the gain sharing methodology where both contractual worker and owner advantage from more effective working by the contractual worker.

Predicting trends and final costs

Predicting trends and final costs is an essential action, as control depends upon the forecast of final costs for everything in the control estimate. Each judgment and forecast should be painstakingly considered by the cost engineer in consultation with the venture organizer.

CONTRACTS / FINANCIAL AUDITS

On fruition of a maintenance work which is at risk to be rehashed, later on, there is an advantage in preparing a post contract survey which should cover the successes and failures of the venture.

Contract documents

An entire survey of the suitability of all parts of these documents should be embraced. The final work scope should be contrasted and that included in the documents and a genuine assessment should be made as to what was precluded from the original which should have been expected. For historical purposes, there can't be excessively detailed in the final work scope.

Any contractual issue which arose during the life of the contract should be noted and preventive measures included in future contracts. In the event that another contract is not required for some time, at that point notes should be connected to the finished contract for reference. Description of priced items should be changed where disputes arose amongst contractual worker and owner.

Invoices and payment

The convenient submission of invoices and their payment should be examined to ensure that sufficient go down information was supplied and that enough time was took into consideration payment. The obvious late payment of invoices can be the cause of the greatest discord amongst owner and contractual worker, albeit some fault can, for the most part, be set on each gathering. The owner will typically dismiss invoices if the following problems are experienced:

- The invoice does not coordinate the affirmed valuation;
- The invoice does not contain the right contractual information, e.g.
- Contract title, contract number, and so forth., or
- The sum is not effectively dispensed to cost centers.

Price of non-conformance

In all aspects of maintenance, the price of nonconformance is considerable to the contractual worker and to the owner both in immediate and indirect costs.

Nonconformance in quality during maintenance has an immediate cost to the contractual worker in amendment costs and usually some consequential costs to the owner in delays, reinsertion, and so on. Nonconformance industrially, e.g. incorrect invoices, late payment of invoices, past due final accounts and the like, make cash stream problems for the contractual worker and sub-contractual workers leading to finance charges and loss of collected interest. (Tonchia, 2008)Some of the alternate items on non-conformance include:

- Late design information;
- Design errors;
- Emergencies (gas alarms, fires, and so on.);
- Late access to items of gear;
- Late conveyance of materials;
- Equipment in worse condition than expected;
- Drawings of existing plant not exceptional;
- Inspection personnel not accessible when required.

These can cause serious delays which may result in disruption to the contractual worker's and the owner's programs and necessitate disaster recovery activity.

In an effort to diminish non-conformance to a minimum, techniques such as aggregate quality management are used by owners and contractual workers to feature, record and amend areas requiring rework and change.

Where non-conformance has been recognized on work, the cost, where possible, should be separated. Future contracts should address the issue if the cause lay with the contractual worker, and should include financial disincentives. The owner's association should survey its procedures if the issue arose in the house.

The genuine requirements should also be assessed as conformance may not be realistically achievable.

CHAPTER 10

Productivity

INTRODUCTION

Productivity depicts different measures of the ability of creation. A productivity measure is communicated as the proportion of yield to inputs utilized as a part of a generation procedure, i.e. yield per unit of info. Productivity is a very important factor underway execution of firms and countries. Expanding national productivity can raise expectations for everyday comforts since all the more genuine pay enhances individuals' capacity to buy products and enterprises, appreciate relaxation, enhance accommodation and instruction and add to social and environmental projects. Productivity development additionally causes organizations to be more profitable. There are a wide range of meanings of productivity and the decision among them relies upon the reason for the productivity estimation as well as information accessibility. (Sakamoto, n.d.)

PURPOSE OF PRODUCTIVITY IMPROVEMENT

Higher standard of living

The general importance of productivity for representatives is enhancing the standard of living by less working hours, higher wages, and more and better jobs, for example. The quantity of normal working hours has reduced in the course of the most recent century however generation volume has expanded consistently after some time everywhere throughout the world. It is straightforward having less working hours in nations with higher productivity than others.

We can state higher wages have been acknowledged in light of productivity improvement. There are no organizations that acknowledge a higher wage level without productivity improvement. The importance of productivity is not just taken after by a number identified with creation, yet in addition the esteem included. (Sakamoto, n.d.)

Job satisfaction

The harmony between workers' physical and mental capacities and necessity for jobs has been highly upgraded by productivity improvement.

Straightforward and very labor-concentrated jobs have been enhanced through motorization, mechanization, and ergonomics (human designing). It is anything but difficult to discover customary labor-concentrated work in factories with out-dated generation strategies or a general undercurrent of lower productivity. In truth, higher abilities are not required for such labor-concentrated obligations since the majority of the work day is for the most part comprised of physical action.

Domination of competition

Worldwide competition is normal today not just in costs of labor, materials, and overhead expenses, yet in addition in the level of productivity. Moving generation work to creating economies is important to management, principally for the advantages it acquires the form of lower labor costs per worker every hour. Be that as it may, is this a correct decision to make? The connection between labor expenses and productivity is appeared by ULCs. For instance, when contrasting creation facilities, if labor costs at one are a large portion of those of the other yet productivity is additionally half, at that point there is no preferred standpoint regarding cost aggressiveness. On the other hand, a plant with double the labor costs yet double the productivity does not endure as far as cost violence. (Sakamoto, n.d.)

There is a straightforward figuring of labor cost viewing productivity as ULC, which was clarified before. ULC demonstrate the development in pay with respect to that of genuine OP. These expenses are ascertained by isolating aggregate labor pay by genuine OP. Changes in ULC can be approximated by subtracting the adjustment in productivity from the change in hourly remuneration. Labor costs, for example, compensation are two times the ULC if the productivity level is half of the normal level. ULC can reduce of genuine wages considerably if productivity is enhanced multiplied, for instance.

Lower consumer prices

You require just contrast 20 years back with now in pricing structure. Because of the same settled specifications yet higher desires among the two representatives and consumers, you can total this up as the consequence of productivity improvement. Lower consumer prices have made it conceivable to enhance living standards extensively without a higher cost trouble. People in created nations as well as those in immature nations can enjoy a superior standard of living due to the consumer items that are delivered in higher productivity nations. Many organizations move their creation plants into low-labor cost territories in immature nations. Does this bode well? Development in itself is not an unsuitable objective, but rather when you take a gander at the decision as far as the ULC point of view, low-labor cost regions don't mean minimal effort on labor charges. It may be even between low labor costs however low productivity and higher labor cost with higher productivity. (Sakamoto, n.d.)

Environmental issues

Nowadays, the earth is one of the key management issues. The objective to consider is not just asset vitality as a segment of items yet additionally over the top WIP and inventory, which can be a misuse of utilizing materials. Another issue is ineffectiveness and inefficiency resulting from pointless work, for example, for WIP and inventory. These things would not be an insignificant measure of consumption spending, but rather the estimation is difficult. The reasoning of just in time (JIT) adds to this issue too. As introduced in a later area, generation following shipment is a simple approach to perceive this condition. (Sakamoto, n.d.)

Effectiveness of capital investment

This is an important issue that management overlooks or has a poor understanding of. All decisions identified with productivity that use capital investment can affect the organization enormously. A capital investment in motorization or automation of human work undertakings alone don't lead to successful improvement of items. Exhibit errands that workers are doing ought to be enhanced before an investment decision is made; something else, that decision won't bring satisfying results. These decisions can lead to lowering profitability while improving productivity. (Sakamoto, n.d.)

We should examine the basic points of consumption and the likelihood to spare use. All things considered, management must have the capacity to measure motivations to invest.

- Necessary specifications of an investment must be examined, especially for an arranged machine, which is a typical investment at manufacturing organizations. Engineers who are in charge of specific errands insist that the new innovation is vital and must be installed quickly. For the situation of a nourishments material maker, the original proposition of consumption was JPY 35,000,000; after close examination, this figure was diminished to JPY 15,000,000.
- A machine's increased capacity immediately was a factor in management's decision, yet deals forecasting was too long range (more than 5 years later on) to justify the investment. Forecasting is simply the possibility of something, and the investment may not be too positive or negative, but rather the more an organization has rehearsed long haul forecasting, the more dependable the forecasted result. Time is profitable too, and this viewpoint ought to likewise be examined. For the situation of a medicine maker, an original investment proposition was 22,000 ton for each year in the generation design. An investigation found a variety of generation capacity

uncertainties on that day, so important errands could be made arrangements for to raise the capacity. Reducing generation misfortune brought about a 3% capacity increase. Another activity was to enhance processing time; the initiative brought about an 8% capacity increase. Because of the improvements, capacity was increased by 20%. All in all, the original arrangement of investment for two vast scale machines was changed to the option of one little machine, an extra to the present machines. Capacity increase was organized at 3,700 tons for every month, and the outcome was 400 under the forecast, or 3,300 tons for each month. This is the reason there was no danger of such a reduction. The first investment design was JPY 1,500,000,000, and it was lessened to JPY 1,000,000,000. (Sakamoto, n.d.)

- Computers' assurance limits are coming soon. For instance, a vendor suggested her organization's adaptation of a software redesign and substitution of current equipment. As told to the consumer, if there is any inconvenience sooner rather than later with the PC and they don't have the most recent form, the organization can't ensure that it will work well under the now-obsolete rendition. There were no issues with the PC before so there is hesitance to get the new form.

In any case, nowadays, it is sometimes less expensive to purchase another PC than invest in one section or one suite of software as a result of these consistent redesigns that the organization invests in. Nonetheless, we realize that substitution timing must be decided by the proprietor, not by the vendor's proposal.

DIFFERENT APPROACHES LEAD TO DIFFERENT RESULTS

Input Reduction First

A sensible approach to manage enough affecting organization performance, for instance, benefit, gainfulness, and efficiency, is to start by decreasing IP resources. The reason is fundamental: IP resources are totally sensible resources inside; there are no requirements from outside of the organization. Start by expanding OP. Any divisions that request expanding IP resources have lost the motivation behind efficiency change in the first place. On the other hand, it is hard to meet administration necessities for an OP increment since it depends upon issues outside of the organization's control. See Figure below.

The second step is utilizing redundant resources for increment OP results. Applying resources is a top administration work. These two phases are tools for focusing on successful efficiency change activity. After utilization of these two phases the result would be that the basic level of efficiency had expanded, and administration had coordinated modify of IP and OP well. IP reduction can be controlled after a set course of action or timetable, however OP increment can't be controlled, and it takes 1 year or more to get results.

Stage one is diminishing IP belonging while at the same time keeping up the present level of OP results. This accepted is important for administration when they plan profitability activity. What does it mean for your organization to target efficiency change thusly? Stage one focuses totally on limiting IP resources for the current OP level. There are a great deal of "essentials" given to workers or wishes of a departmental supervisor to build OP. Nonetheless, an important point to consider in the start of any profitability matter is to focus inward organization approach on diminishing of benefits. The reason is that the focal exercises are not enormous issues and the basic terms to complete this stage don't take a long time. Resource control is a possible control for any organization. (Sakamoto, n.d.)

Remember diminishing working hours, number of workers/operators, and inadequate work endeavors wind up clearly basic things because of external factors. The start of profitability increment should be started at this stage. (Sakamoto, n.d.)

Stage two is utilizing abundance resources for current OP necessities. Profitability change never advocates terminating people; the legitimization is to keep alter between reduced OP and required IP. Shutting particular factories is an instance of this. To discover new business scopes of development, reevaluate purchasing choices and lessening thing costs in order to get more arrangements orders. On salaried positions, for instance, inside the building division, profitability change is possible through assigning specialists to locate these new business zones of development.

Stage three is overseeing two parameters of efficiency that depend upon organizations' necessities. A larger amount of adjusting between OP conditions and the necessities of IP resources is essential. What measure of efficiency change in a year is a target for organizations and why? No under 5% consistently is the base most negligible target, yet 1% consistently (no under 10% consistently) is endorsed. Extraordinary organizations increment their efficiency by more than 20 to 25% consistently for over a maintained time. Yearly change of 10% makes 160% inside 5 years, 20% makes 200% inside 4 years and 25% makes 200% inside 3 years. (Sakamoto, n.d.)

Advantages of efficiency development

Efficiency development is an essential wellspring of development in expectations for everyday comforts. Profitability development infers more regard is incorporated creation and this infers more pay is available to be passed on.

At a firm or industry level, the advantages of efficiency development can be coursed in different ways:

- To the workforce through better wages and conditions;
- To investors and superannuation finances through expanded benefits and benefit dispersals;
- To customers through lower costs;
- To the earth through more stringent natural protection; and
- To governments through increments in control portions (which can be used to support social and natural tasks).

Profitability development is important to the firm since it infers that it can meet its (perhaps developing) responsibilities regarding workers, investors, and governments (costs and control), and still stay engaged or even improve its forcefulness in the business focus. Including more inputs won't build the pay earned per unit of input (unless there are expanding returns to scale). Without a doubt, it is most likely going to mean lower normal wages and lower rates of benefit. Regardless, when there is profitability development, even the current duty of benefits makes more yield and pay. Pay made per unit of input increments. Additional benefits are in like manner maneuvered in into creation and can be productively used.

DRIVERS OF PROFITABILITY DEVELOPMENT

In the most incite sense, efficiency is dictated by the available advancement or know-how for changing over resources into yields, and the way in which resources are organized to convey stock and ventures. Historically, efficiency has upgraded through development as procedures with poor profitability performance are abandoned and more avant-garde forms are manhandled. Process changes may incorporate organizational structures (e.g. core capacities and supplier relationships), administration frameworks, work strategies, fabricating techniques, and changing business sector structure. A prestigious case is the mechanical generation framework and the procedure of huge scale producing that appeared in the decade following business presentation of the automobile. (Sakamoto, n.d.)

Expansive scale fabricating fundamentally diminished the labor in creating parts for and collecting the automobile, however after its no matter how you look at it adoption profitability picks up in automobile era were much lower. A similar case was seen with jar, which saw the most noteworthy efficiency picks up in the early decades after presentation. Various different ventures demonstrate tantamount cases. The case was again trailed by the PC, information and correspondences enterprises in the late 1990s when an extraordinary piece of the national profitability picks up occurred in these ventures.

There is a general comprehension of the fundamental determinants or drivers of profitability development. Certain factors are fundamental for deciding profitability development. The Office for National Statistics (UK) recognizes five drivers that connect to underlie whole deal profitability performance: speculation, development, aptitudes, wander and rivalry. (ONS 3, 20)

- Investment is in physical capital — apparatus, equipment and structures. The more capital workers have accessible to them, all things considered the better they can do their employments, delivering more and better quality yield.

- Innovation is the productive manhandle of new contemplations. New musings can show up as new progressions, new things or new corporate structures and techniques for working. Accelerating the scattering of advancements can support efficiency.
- Skills are characterized as the sum and quality of labor of different sorts open in an economy. Capacities supplement physical capital, and are relied upon to abuse interest in new advances and organizational structures.
- Enterprise is characterized as the seizing of new business opportunities by both new organizations and existing firms. New endeavors match existing firms by new contemplations and advances expanding rivalry. Specialists can join factors of era and new advances forcing existing firms to modify or leave the market.
- Competition upgrades profitability by making motivations to improve and ensures that advantages are apportioned to the most proficient firms. It moreover forces existing firms to organize work more effectively through pantomimes of organizational structures and advancement.

ENGINEERING APPROACH FOR PRODUCTIVITY

Numerous administration groups have executed certain techniques for development with in mode systems, for example, QCC, kanban, and Kaizen. Organizations have demanded that best administration touted the consequences of different organizations. Regardless of the possibility that restricted is a successful system in an organization, it is hard to state the strategy is the most ideal path for another organization. The choice is generally in view of confirmation at shops, in any case, once more, they may not on time convincing outcomes with respect to productivity organization wide. It is anything but difficult to discover fizzled cases of a method that was actualized essentially on the grounds that it was stylish. To characterize the strategy's motivation and proposed explanations behind being executed is the primary need, and afterward diagram the imaginable comes about. (Sakamoto, n.d.)

Required by administration, the approach includes growing new generation strategies that can address the issues of administrators who need to maintain a more lean business. i.e., it alludes to estimation to accomplish supervisors' desires and prerequisites for cost decreases and productivity upgrades, regularly high focuses on that are hard to accomplish with the above popular methodologies.

"Target approach" on timely entertainment of results or keep up a thin variety of results, As depicted before, the inverse of target approach is subjective approach, which offers a ton of contrasts or a high level of variety on comes about.

One subjective approach is an "arbitrary approach", which makes enhancements in a steady progression with whatever thoughts you have. Those thoughts in this approach of change might be successful at one time, yet steady sufficiency is doubtful what's more, the approach needs encourage examination. For instance, you can get one of those widespread change methods and execute it without satisfactory examination of its general adequacy. This brings up an issue about the activity also, capacity of the administrator / managers. What is the most ideal approach to oversee issues and issues in our business? Any occupations in view of a practical association ought to accept a "target arranged approach" instead of a "systems situated approach". (Sakamoto, n.d.)

Be that as it may, business hones generally rely upon a subjective way to deal with choice making. This book is not proclaiming that is correct or off-base. By and large, it is not an distortion to state that there is no perfect way or administration style. To whole up, it is very important to empower the capacity of not a private expert but rather an hierarchical expert and to accumulate diverse perspectives and conclusions of individuals' association wide. You should remember a mission when dealing with your undertaking. You should do your best on your employment, as well as watch out for other offices and the organization as a whole. Diverse Methodologies Prompt Distinctive Outcomes.

A typical approach is to visit a shop and discover issues and after that move forward them now and again. Some trust that the expert who adopts this strategy is to be respected, however in this approach the shop floor is not taken a gander at as a framework. A "precise approach" is a clear and composed method for development; it takes a gander at the objective in light of a wide and high perspective. It is additionally called an "issues arranged approach". While the arbitrary approach can bring about a speedy arrangement, the orderly approach appears to take additional time. (Sakamoto, n.d.)

At the end of the day, the irregular approach is a sensible or potentially subjective method for accomplishing upgrades and discovering arrangements. It is likewise called a "method situated approach". Given all these methodologies, in the event that you are not a specialist on change, you should to pick solid quality over speediness. To put it another way, approach should to be more similar to a wide plane than an expansive dab or a thick line. Execution of the deliberate approach can accomplish numerous noteworthy outcomes that are specifically connected to corporate execution. Besides, the building based approach, MDC, for instance, is a practical approach for such purposes as getting a legal headcount underway, limit increment of significant machines and equipment, and setup time lessening. The two methodologies for enhancing working techniques were specified before, yet how about we sum up for an alternate reason in light of the fact that a important issue among administration is not setting aside the opportunity to know the distinction in methodologies, for example, "arbitrary approach" and "methodical approach". Since there are considerable measures of tasks in shops that administration needs to manage well in regards to enhancing productivity and cost diminishment, it illuminates the methodologies much further. (Sakamoto, n.d.)

One approach is specialized IE and the other is building IE. Inside the suite of modern products, engineers' thoughts will probably transform into design designs if the designer has demonstrated process-arranging background. Process design relies upon a particular sort of administration worried about procedures, the design may thrive because of worthy procedures. Does this mean any plans are pointless?

Architects who speak to those projects with a design target typically have a foundation in view of one of the fields of hypothetical designing. One such field could be mechanical engineering. Mechanical engineers design process designs that incorporate generation techniques, strategies, and the design of a plant. Engineering thought all in all is not only to think about frail focuses, imperfections of products, or, on the other hand deformities of the generation procedure. They consider economy and rearranged techniques, and afterward set those musings into standard operation methodology. (Sakamoto, n.d.)

THREE LEVELS OF IMPROVEMENT

By and large talking, there are three ways to deal with enhancing creation strategies (see Figure 4.5). These are: taking out waste or work disentanglement, making enhancements, and improving. The primary approach includes expelling waste (muda) found by watching conditions on the plant floor, as found in little gathering movement (SGA), Kaizen, and other self-starter exercises, and in this manner known as work improvement.

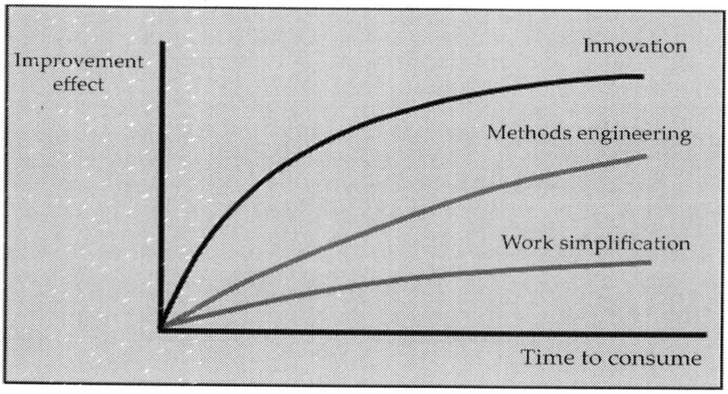

Figure 101: Improvement spread (Sakamoto, n.d.)

The second approach involves expanding productivity levels despite when there is no loss by taking convincing thoughts gained from different organizations, for instance, and specifically presenting those systems and techniques. Diminishments in retooling and set-up times accomplished by many organizations using single minutes of set-up are a case of this approach. The third approach includes growing new generation strategies that can address the issues of directors who need to maintain a more lean business, i.e., it alludes to measures to accomplish supervisors' desires and necessities for cost decreases and productivity changes, regularly high focuses on that are hard to accomplish with the initial two methodologies. (Sakamoto, n.d.)

Regardless of the possibility that all industrial facility floor waste (as a matter of fact hard to characterize, since what matters as waste might be subjective) is dispensed with, it offers no certifications over finding new generation techniques that can give an organization upper hands. So also, regardless of the possibility that compelling methods from different organizations are effectively presented, it doesn't diminish the longing of the organization to accomplish more elevated amounts of productivity than those of the organization that created them. The improvement of convincing generation strategies that can address administration issues is something that can be accomplished by designing new creation strategies.

When searching for approaches to enhance productivity by designing new generation techniques that will directly affect comes about, it is vital to choose which of the three ways to deal with take. To clarify the contrasts between the methodologies, let us consider a straightforward case: imagine a full moon. How OK characterize it? What I need you to do here is imagine how you would show what sort of a thing it is to someone else.

"Distinctive methodologies" could closely resemble taking a gander at it with the exposed eye, seeing it through a telescope, or sending a carry to arrive on the solar surface. With the naked eye, the moon is sentimental and resembles a major ball; through a telescope the numerous holes come into view; and arriving on the moon a telescope the numerous cavities come into view; and arriving on the moon implies that you will have the capacity to see only the destroy solar surface. Along these lines, taking diverse methodologies brings about totally extraordinary perspectives or meanings of "full moon". Essentially, adopting the three unique strategies to generation techniques can give you infinitely extraordinary perspectives; i.e., change what you see. (Sakamoto, n.d.)

In the event that the administration is firmly inspired by usage of every one of those wellknown systems, they can't rival the other great standing organizations who more likely than not proceeded. Such administration may wind up as serious administration of cost diminishment that does not permit even small waste. When I say waste end, change, and development, waste end alludes to a lessening of waste and bother, while change alludes to a copy execution of contenders' great illustrations. These are not anticipated that would prompt a complete administration approach covering an interesting creation framework and its related issues. Any organization requires a key administration approach that is expressly aware of the development level from the earliest starting point, regardless of what the current level is. A straightforward case of change and advancement is word handling by (PC) today. The manual was enhanced: the ink was made darker, the machine itself was made lighter, and the ball component was presented, and so on. Word processors, then again, were an advancement which had been created after word processing programming for PC. Development in individual PC design proceeds with every year.

Three phases are included in growing more successful working strategies. You can go to any shop floor and discover some wastefulness. It might be moved forward instantly much of the time, and execution is not troublesome. Different sorts of change must be created, in any case, to accomplish the maximum capacity for development. They might be grouped into the accompanying classifications:

work improvement;

Techniques building; and

Advancement

Three techniques for change exercises are abridged in Figure 4.5. The most critical contrasts between the three exercises are diminished nonworking time or potentially AF work or expanded BF work. It looks as though they are comparative, however real outcomes accomplished from each of them are entirely unexpected.

Work Simplification

This sort of change, which is the most minimal level of change, is regularly concentrated on taking out pointless work or limiting any wasteful aspects required in an operation in light of subjective separation that it is waste, for example, wasteful work substance. SGAs, for example, quality circles and the examination of current strategies or procedures all have a place in this class. While this range gives an amazing chance to accomplish comes about through an expansive number of differed thoughts, just a little rate of the collective change potential will be accomplished through these methods. Regularly, these enhancements don't require real changes in existing machines, offices, or formats and subsequently don't include a lot of capital consumption. (Sakamoto, n.d.)

The initial phase in work separating incorporates recommendation designs, self-rule circles, and waste change when it is found in shops by laborers themselves. It is successful to share productivity issues between representatives (engineers) and administration. Yet, the outcomes won't not have a substantial effect on corporate execution in general.

When strategies have been enhanced through work improvement, it might be felt that the subsequent working practices should to be equivalent with different organizations or demonstrated administration procedures. Be that as it may, this may not be the situation, since fluctuating levels of work improvement potential exist in various circumstances. So as to coordinate or exceed the productivity of different frameworks that are being utilized as a standard for worthy productivity levels, other change systems must be utilized.

Work rearrangements is recently worried about waste; change is not as restricted as work improvement but rather executes duplicating others' prosperity illustrations such as Kaizen and Toyota creation framework (TPS). All things considered, they may never make up for lost time and exceed the firsts regardless of the possibility that they took after their cases for a long time.

The techniques building classification will require minor changes to be made to creation equipment, for instance, change to work environment design, installations, machines, and tooling. This frequently requires a little measure of speculation, which might be effortlessly advocated.

Methods Engineering/Improvement

The second approach is enhancing current strategies through modern designing instruments with regular or conventional methods in regards to productivity issues. It has been known as taking "logical strides" for taking care of issues. The purpose of this approach is a point by point investigation of current strategies in rehearses. What occurs at shops? A curious demeanor is very important for this strategy as inquiries may come about in finding new answers for productivity change. It may get more powerful comes about than the waste disposal approach, however an inquiry might be excessively essential for current techniques. What is the importance of current strategies?

At a forceful level among contenders, the amount of change is found or, on the other hand it will not be a critical point. There is an objective that an organization should go after, which is a level superior to "aggressiveness". Those objectives may not be achievable utilizing current strategies. At the end of the day, truly practical arrangements may be out of the extent of specific classifications of enhancements. (Sakamoto, n.d.)

A compelling change method created by Shigeo Shingo was named the single moment trade kick the bucket approach (SMED). This approach had an objective setup time of less than 10 min. Many organizations decreased their setup time at their compress shop, for instance, to under 10 min. Nonetheless, just the principal organization, who built up the kick the bucket setting thought, had a genuine upper hand over the different organizations, who were a few years behind the first thoughts. To gain important ground, in this manner, we should build up another state of mind concerning strategies change.

This approach expects time to be spent in examination of the present circumstance, to recognize issue zones, and to apply presence of mind to clean out the waste and poor practices by growing new strategies. Conventional modern building preparing puts the stress on the above focuses, and most mechanical designing instructional classes and books are about change of the present work strategies, depending upon little gathering projects. The benefit of "work rearrangements" is that the cost and time required to apply these techniques are moderately little. The shortcoming of this approach is that it doesn't concentrate on the significant open doors for enhancing productivity.

Innovation

The third approach is a more standard idea today and critical in the present extreme cost aggressiveness conditions. We should leave past techniques behind (those that did not have a demonstrated reputation) for finding powerful arrangements that meet future conditions well. There are three focuses for practical deduction for achieving creative levels however much as could be expected. One is to characterize working models rather than current strategies, consider works as the premise of techniques, for example, operations or component operations, and after that set targets in light of objective and sensible parameters. General ways to deal with designing consider this start. It is straightforward the distinction amongst change an development of a product. How about we take a gander at a case:

There are two helpful introduction gadgets: the 35-mm slide projector and the overhead projector (OHP). Utilizing a slide projector is the conventional method for making introductions. Some advanced elements have been included, for example, remote control, forward, in reverse, centering, and cartridges. These sorts of changes have made it simple to utilize the slide projector for speakers at introductions, for instance. however, there is as yet a bother with the slide projector: the fundamental switch.

One can turn a slide projector on and off just on its shell and not at the speaker's position. It is anything but difficult to control the ace switch at the speaker's position if you introduce a switch. This sort of change is named work disentanglement or potentially techniques building. (Sakamoto, n.d.)

Then again, there is the OHP. The OHP is effortlessly killed and on at the speaker's position, and every single other purpose of bother with slide projectors are overcome. The inquiry is whether the OHP was created as a change on the slide projector. The appropriate response is no. The OHP was created without progressing the slide projector. The particular of overhead projection is characterized and designs built up the OHP as opposed to making an amassing of consistent little changes to slide projectors. This is a case of advancement of products furthermore, product design.

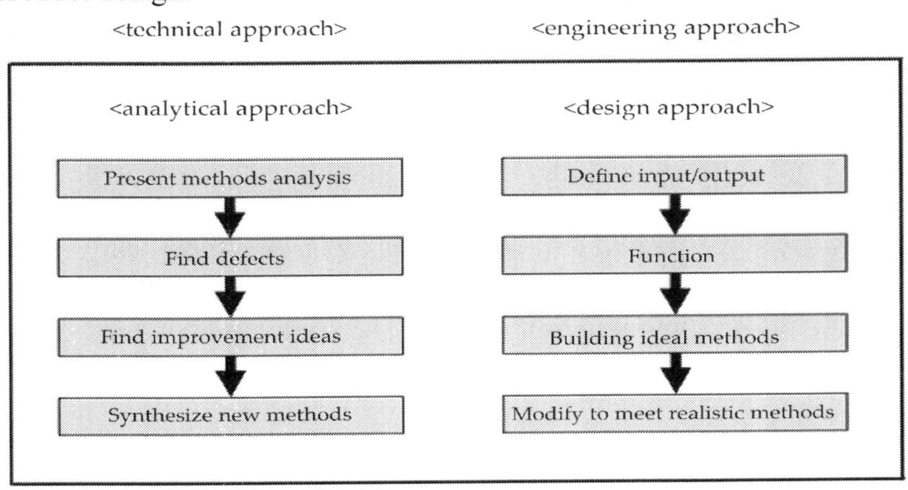

Figure 102: Design approaches (Sakamoto, n.d.)

Another case of advancement is the air-controlled nailing machine, which is sustained by a magazine or cartridge of nails. The regular way to deal with moving forward the nailing of wooden structures is to create enhanced strategies by concentrate the current operation. The operation is then enhanced by rearranging the getting a handle on and situating of a nail and by trying different things with various lengths of sledge handles, pound head shapes and weights until the point when the ideal combine is obtained. This takes after the strategy Taylor followed in his celebrated measure tests. It is the traditional techniques designing methodology.

The advancement approach is to consider the BFs of the operation, to build up a totally new strategy for performing them without respect to the present technique, and to achieve the BF in at least time. This is the approach that would lead to the advancement of a nailing machine to supplant a customary mallet and nail operation. With already depicted sorts of techniques change, you may have a vast number of little thoughts that can just take you up until now. The approach must be changed, or productive and innovative thoughts won't be found to raise the potential for ability. This level is accomplished through the design approach, for example, the MDC. (Sakamoto, n.d.)

A change target must be set as high as could reasonably be expected and by the change extend itself as opposed to through the replicating of other tasks' achievements. The MDC may not be connected to current techniques since imaginative upgrades do not work through examination of current techniques. Set up of this approach, appropriate strategies for characterized Operation ought to be made with the goal that development comes about. Having gone through the MDC, advancement will achieve the perfect level. Work rearrangements and strategies change are both diagnostic methodologies or, on the other hand specialized approach; development is a design or building approach. When you utilize customary techniques change, for example, work disentanglement, you will be keen on the separation for getting and putting parts, using the two hands, and enhancements without changing primary parts of dances and installations. In the event that you take after the MDC, be that as it may, there is greater chance to find a very surprising method for enhancing strategies that don't simply take after the present, essential operation techniques. (Sakamoto, n.d.)

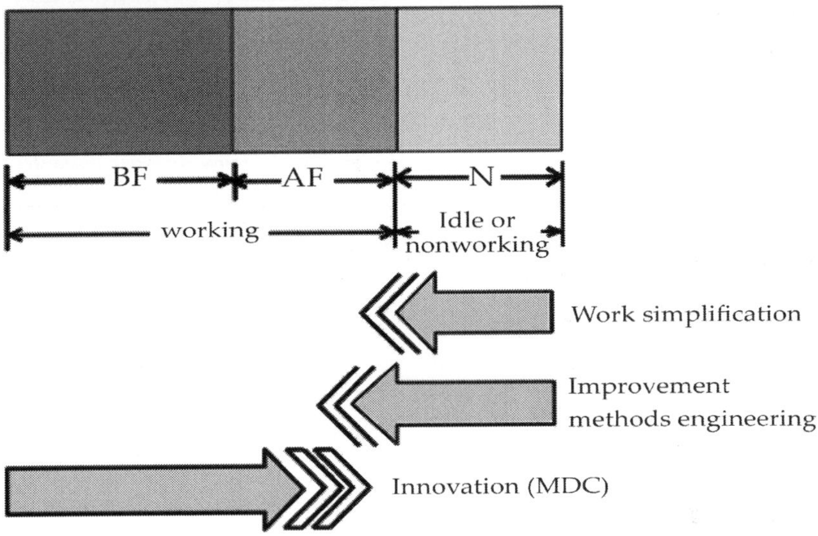

Figure 103: uprising BF ratio (Sakamoto, n.d.)

The purpose of this distinction is whether you focus on the BF or not. This is the reason the MDC will prompt a hop to another, higher stage. See above two Figures. Inside assembling, the operations and procedures are methods for changing over from an IP status to an Operation status. At a lower level are operational components and movements, some of which contribute straightforwardly to taking care of Operation requests and others which boost preceding upstream work substance. These two classes have been characterized as BF and AF. The most huge contrasts between the three exercises are diminished nonworking time and additionally AF work or expanded BF work. It looks as though they are comparative, yet genuine outcomes accomplished from each of them are entirely unexpected. The MDC builds BF work for higher productivity change and creative thoughts.

"Advancements require a combination of numerous thoughts to succeed, including new products or administrations, empowering advances or abilities, boundaries to section from contenders, a convincing plan of action, and fundamental organizations" (Carlson and Wilmot 2006).

There is a straightforward case of assembling process development: the mini-computers. It utilized be get together work with screws and nuts, however today it is delivered in the process industry without fasteners and nuts. This is an effective instance of products and process development. You can't ever discover arrangements of genuine larger amount of productivity with Kaizen. In other words, productivity methodology should change from correlation with the past so as to achieve higher targets, yet this can't occur without a completely reasonable emotionally supportive network from administration. The inquiry is: The thing that productivity target is proper for the corporate level of violence?

THREE DIMENSIONS OF PRODUCTIVITY

This section follows the principle that you take after different routes in comparison to you did in the past. Diverse approaches find distinctive solutions for productivity. What amount of improvement do you require as the objective? Theoretical order of productivity methodology in view of industrial engineering is required. With a powerful engineering approach, it is possible to guarantee the achievement of a sensible level of results. To achieve a more elevated amount of productivity improvement, there are two keys to success. One is to separate productivity substance into M, P, and U, which depend on industrial engineering techniques. Another is to adopt new approaches to each dimension of productivity. M, P, and U encompass total productivity improvement. (Sakamoto, n.d.)

Points of Successful Productivity

Examples of remarkable productivity improvement by Japanese manufacturers reign in the previous figures. The two companies demonstrated are producers of sustenance and nourishment materials, such as ham, sausages, and powder materials. What they accomplished for the results included improved production methods in shop floors for reducing number of specialists with MDC and work performance control with designed standard time. They are perceived as the overall model for working pace standard.

There are three important points to productivity success. The most important one is "changing approach for productivity matters", which means you can get truly unique results on the off chance that you take after unexpected courses in comparison to you did in the past. Distinctive approaches find diverse solutions for productivity. An adjustment in real life prompts an adjustment in results. The second point is that you should make your own particular manners as opposed to copy others' experiences. You might increase some success through copying but it will probably be here and now. Once more, you may achieve a level similar to the first company's level but the approach does not prompt anything further. Focus on making a solution for what you need, not how to simply make a copy.

How much improvement you require as the objective can't be found through after popular methodologies. The third point is: theoretical order of productivity depends on targets and methodology in view of industrial engineering. These tools would be the three dimensions of M, P and U. Industrial engineering provides enough experience and background of hypothesis to make enduring results. The engineering state of mind is more important than just getting results without such a background. (Sakamoto, n.d.)

Edward V. Krick stated: "Engineering is primarily worried about application of expository methods, principles of physical and sociologies, and the imaginative process, to the problem of changing over our new materials and others resources to forms that fulfill the need of mankind.

The process required in taking care of the transformation problems is conventionally alluded to as outline" (Krick 1965).

The engineering approach or state of mind contains audit of got results and systematic steps to get those results in any case. It is not an appropriate approach for those with massive ability or a considerable measure of shots. With a viable engineering approach, it is possible to guarantee the attainment of a sensible level of results without broad experience.

Relationship of M, P, and U to Standard Time

Step by step progression of phases is important for success from the perspective of company performance. There are three possible phases. It is important that management wishes to achieve higher productivity. To achieve a larger amount of productivity improvement, there are two keys to success. One is to separate productivity substance into M, P and U with dependence on an industrial engineering techniques point of view. Another is to adopt new approaches to each dimension or function of productivity. Those three dimensions multiply productivity results. The

M dimension contributes adequacy and the P dimension contributes effectiveness. The U dimension can't provide clear results without these two dimensions. Collaboration of the three dimensions for improvement is ultimately the most successful objective. Consider a simple example of travel time dependent on activity methods that have awesome potential for the future (Table 5.1, Helmrich 2003). For go of about

Table 10.1 M, P, U and travel time (Sakamoto, n.d.)

	Car	Train	Airplane	Potential
Method	10h	3h	50min	300%
Performance	9-11h	3-4h	50-70min	20-30%
Utilization	Disturbances	Minor delay	Delays normal	20-30%
	To city	To city	Connections	

500 km, an auto takes 10 h, a bullet prepare is 3 h, and an airplane takes 50 min. This comparison is just about equipment such as auto, prepare, and airplane; this is the M dimension of productivity. Nevertheless, there is a software issue for each travel method also. Above is the scheduled amount of time, but actual time may fluctuate 9–1 h for the auto; the prepare may be postponed 1h; the airplane can differ much more; this is the P dimension. Some other time misfortune may happen due to the venture plan; this is the U dimension of productivity. Total travel hours are dependent on these three dimensions. Airplanes means high-cost machines, but it doesn't guarantee an abnormal state of productivity.

What is exciting about the factors M, P, and U is that together, through multiplication, they give the magnitude of the total productivity improvement. Take a stab at calculating the total productivity improvement in the example given underneath. Begin by calculating the relative improvements of M, P, and U, and then multiply the factors to produce the total productivity improvement. Following a simple calculation demonstrates the usefulness of the three dimensions' collaboration, for example:

$$M \times P \times U = 0.8 \times 0.8 \times 1.0 = 0.64 \text{ case A}$$
$$= 1.2 \times 1.2 \times 1.0 = 1.44 \text{ case B}$$

1.0 means normal 100% and normal level compared to the industry normal. The U dimension is set at just 1.0 because U does not have a high potential of productivity improvement. Case A is 0.8 for the M and P dimensions; that means a tad bit lower than the industry normal. It is a condition in which with just a tiny bit of poor working methods and/or old machines are applied as working methods and a taste of poor levels of P, the performance level generally speaking is at a lower level compared to the standard. That 0.2 (20%) of lower level is not enormous enough for either dimension M or P but the result of multiplied M and P prompts 0.64, almost a half level behind for industry normal productivity. The results are too a long ways behind the standard, which is not an acceptable level. A question is, do you measure productivity in these three dimensions?

Case B indicates 1.44, or a 144% level of productivity compared to the industry normal. Just a 20% expansion for M, and P results in a 44% expansion for total productivity. For the most part, 200% (two times) or 400% (four times) improvement of productivity may not be a trustworthy objective, but rather a 150% improvement for both of M and P prompts a 225% improvement, more than two times of productivity improvement. (Sakamoto, n.d.)

Dimension of Methods

M is, without comparison, the most dominant dimension as to improving productivity. All around applied methods likewise make motivation and better U, since the process becomes more secure (less disruptions). It is therefore important to begin with methods development while sorting out productivity improvement. Methods can be isolated into two classes.

The first is equipment, such as machines, tools, layout, and the like. Management simply chooses to implement new machines or automation thoughts for an expansion in productivity or to spare the number of specialists, but it requires a high amount of expenditure. A simple approach to apply this productivity improvement begins with the premise that there is no prior competitiveness among competitors because they have not adopted those new machines from outside suppliers. In my experience, I never refused those investments if a more elevated amount of productivity could be proven with this specialized development. But management has too often taken a mild-mannered position when managing coveted OP for their company. Sometimes those thoughts to purchase machines are less important than different thoughts, such as the proper preparing of employees.

Anyone schemes quickly on the off chance that they take airplanes as opposed to trains. There is a tram system called Shinkansen in Japan. The maximum speed is 300 km/h. There have been no mishaps since the Shinkansen began business in 1964. Trains leave each 10–15 min from terminal stations. Shinkansen is the most advantageous prepare in Japan, just as specific factories have the most modern offices of manufacturing. But anyone can take Shinkansen on the off chance that they like. This means absolute level of travel time is improved but the relative level among voyagers is still even; there is no advancement on travel time.

There is a similar sort of misunderstanding made day by day in regards to productivity issues. Companies trust they have achieved a larger amount of productivity by implementing modern, cutting edge offices, but no modern office all by itself can guarantee a competitive level of productivity. (Sakamoto, n.d.)

The second class is software, such as motion patterns, association, preparing, and support association. There is often a lack of method on the software side. The risk of limiting improvements to this territory is that the enduring impacts are limited. As a result, a positive and/or negative impact will be found.

An impact of prolonged focus on the zones where improvements were made is the unwillingness to discover that improvements won't keep going forever without sustainable management and dimensions of productivity.

We should investigate an example of managing hardships on line adjusting through software. Adjusting misfortunes can be partitioned into two classes: SLB and DLB. SLB manages a careful adjusting of the capacity of various stations along the production line by industrial architects. DLB, which is more dominant recently, is significantly more difficult to handle because of varieties in the work substance of various products that are manufactured on the assembly line; it is difficult to rebalance the line. Natural quality varieties of IP material and specialists' performance variety of time in cycles both have a place with the DLB impact. One approach to manage dynamic adjusting misfortunes is teamwork on the production line, where employees are encouraged to readjust the line by designating errands and resources at each work station. All together for the employees to adjust to this adaptability, they must be multi-talented and more anxious to step up with regards to specific circumstances.

There is often a lack of method deal with the software side. The risk of limiting improvements to this zone is that the enduring impacts are limited. An impact is produced insofar as a focus is maintained on the ranges where improvements have been made. But, the enduring impacts don't come automatically.

Changes to the equipment often prompt dependable impacts once the equipment is in place. A combination of both the equipment and software improvements is the way to success and enduring impacts.

It is possible to isolate software into three zones: manufacturing system, manufacturing methods, and management system. Another point we must pay regard for is the dimension of M where genuine competitiveness happens when the other company has propelled software instead of equipment. Equipment domination requires perceiving what has happened with competitors. Meanwhile, software advancement of manufacturing is never open to different competitors. Information identified with any inner changes of the software condition should never be unveiled remotely, especially to your competitors. This can't be perceived without going by and watching their shop floors. (Sakamoto, n.d.)

General improvement issues of M are:

- BF time should be the maximum amount of process duration.
- Movement elements of machines should be parallel to move simultaneously.
- Moving separation of unnecessary movement should be minimized (especially empty movement).
- Work stations' process duration should be sorted as DLB as opposed to SLB.
- Set moving speed of machines at maximum possible of coherent and sensible speed.

Dimension of Performance

P, i.e., the motivation of the employees or the speed of the machines, is often emphasized as the means to expand productivity. P is important, not the minimum since it produces a multiplier impact on the improvement of M and U. It must, be that as it may, never be the main measure for improving productivity. Telling employees how important their function contributions are, combined with clear targets creates motivation.

The speed of machines often changes incredibly. There is a considerable measure of potential in reducing these varieties and the possibility of raising the speed to that claimed by the manufacturer. Low machine speeds are often due to specialized inadequacies or disruptions. It is just once these have been eliminated that speeds can be brought up in a controlled manner.

P is the result of measuring and comparing amongst standard and actual time. Standard time is usually in light of the standard methods such as measurement of laborers' actual P compared to the standard working methods. Standard time is characterized by "given" methods. The most important part of the definition is in regards to standard methods. Actual working time depends on the applied actual methods as opposed to an individual laborer's pace. There is a simple misunderstanding about measuring specialists' P with working pace or speed, for example. There is a meaning of working pace that says standard pace is the meaning of standard time.

Practices of measuring performance and improving the level of P are primarily due to following standard working methods themselves. Laborers disregard the standard methods and supervisors don't understand the need of supervising instruction of the standard working methods to every specialist at their shops. (Sakamoto, n.d.)

Another point is to disregard the worldwide standard of working pace such as assignment standard, which has been approved as the practice around the world. A practical tool to know this overall standard pace is through application of MTM. The measured results where standard time is not applied normally show a low level of P such as around 60% for a standard of 100%. This means an almost 170% improvement potential on performance. Why do you miss the potential without thinking about it?

There is no purpose behind management to disapprove standard time. There is a misunderstanding that management dislikes to introduce standard time and/or MTM to their shop floors. Standard time require just be set and appeared to laborers, alongside any supervising activities such as instruction of standard methods. There are likewise similar issues concerning both the machines and processing time, because there is no world standard like MTM for them except machine tool operations. Be that as it may, the distinction of standard and actual time is effectively found in practices due to an absence of hypothetical background with respect to current machines and processing time. As per practical experience, the performance level of machines and processing time is a low 80% compared to sensible and theoretical standards of machines and processing time. (Sakamoto, n.d.)

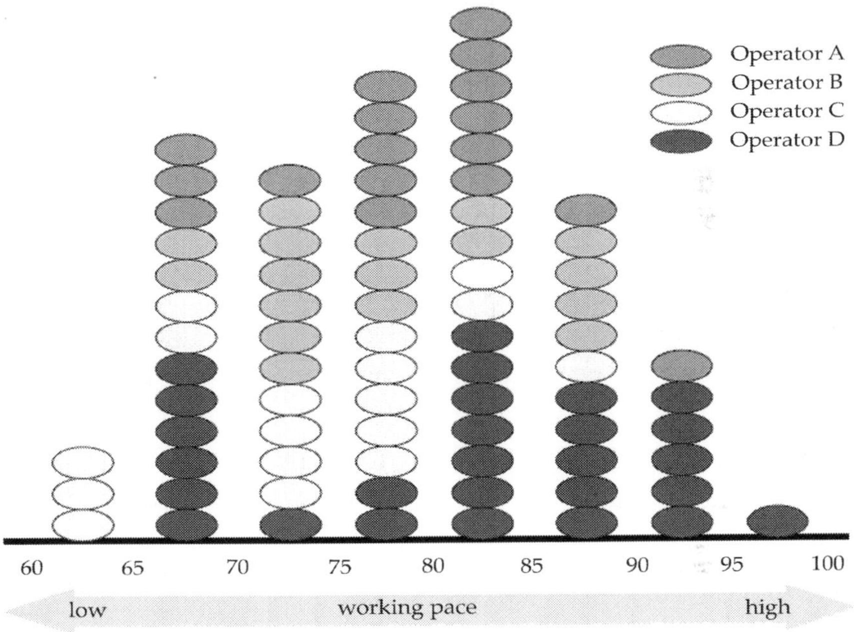

Figure 104 DTS result on an assembly (Sakamoto, n.d.)

Compared to a standard of 100%, more than 130% of improvable potential is there with respect to the performance level of machines and processing time. None of these figures should be overlooked by management although it is small. The improvable potential of the dimension of P is up to standards itself; industrial designers are expected to perform at their maximum even at the larger amount of standard time. Practice of fluctuation time value is appeared in Figure above.

General improvement issues of P are:

Complete division of employment functions among operators and other functional associations.

- Communicate specifically between two movements.
- Measure P compared to build standard time and improvement points.
- The specialists should be prepared with the most ideal way that is standard working methods.
- Supervisors should remain on their shop floors a maximum amount of move time to watch and instruct their specialists.

Dimension of Utilization

U is an expression, for the two machines and humans, of the proportion of planned time that is used for exercises and makes value. Common troubles for the two humans and machines are often caused by specialized unsteadiness. This can be a result of the poor condition of equipment, low material supply, and different various changes in the quality of components. Under an over the top amount of merchandise stream, adjusting misfortunes often occur, because the workload was not equally distributed to all the work stations in the work process. Be that as it may, varieties in the work substance of products generally complicate the matter much further. (Sakamoto, n.d.)

In intensified produce stream, adjusting adversities often emerge, such as difficulties in distributing work equally between all work stations in the stream. Vast varieties in the work substance of products further complicate the matter. Production planning and control, office maintenance, and quality control are exercises identified with the dimension of U in productivity. Those exercises in regards to U by and large are the responsibility of support staff but it is as yet possible for those contributions to productivity to be measured against sensible standard time.

U contribution to total productivity does not have a bigger potential compared to M and P. In any case, one thing to remember is that the level of inventory and/or WIP, does not make any impact from the productivity point of view. This means strife between partial productivity in total plants and expansive productivity can emerge because process duration reduction just happens at a specific area, department, or team. That means genuine contributions to productivity should be expanding the total level of productivity as opposed to specific shops or work stations. The process must be association wide. How would we utilize a partial level of productivity to form the contribution of U?

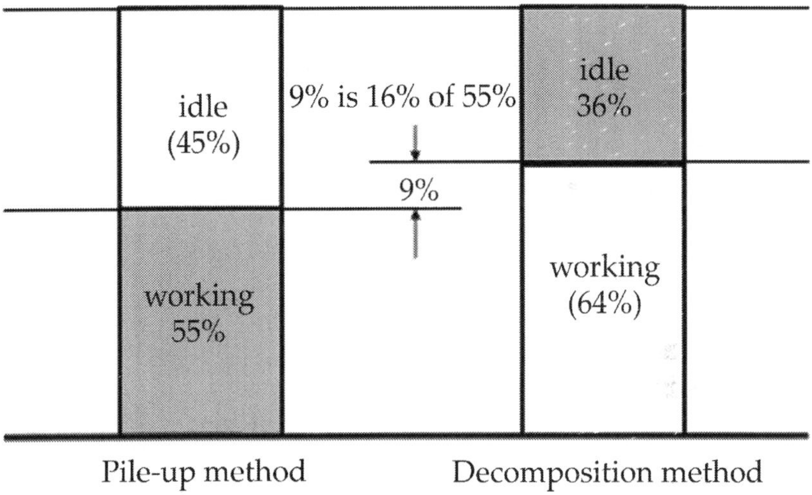

Figure 105 pile-up and decomposition methods (Sakamoto, n.d.)

There are two fascinating approaches with respect to U. Those are the "pile-up method" and the "decomposition method" of U. U percentage is normally measured as reported sit still or nonworking time subtracted from the base; this means one move minus reported nonworking time is U. This is the "decomposition method" to measure U: (100 – 36) = 64%. Another approach is to calculate U as per standard time. Total standard time as the amount of completed work is working time; if working pace is 100%, at that point 55% of U can be accomplished using the piling-up method, the remaining 45% is nonworking or sit out of gear time. (Sakamoto, n.d.)

Figure above demonstrates a comparison between two methods of measuring U. Working time is 55% for the pile-up method but 64% for the decomposition method. The distinction of the two measuring methods is not small; 9% and 9% is 16% of 55%. A distinction of more than 20% is not unusual as indicated by practices. This distinction is an issue of the U function. Figure presented immediately below is called "accumulated chart". The vertical stem is the quantity of volume in regards to production, shipping, and inventory; the level stem is production days. The distinction of volume between production an or B and shipment is inventory. The chart makes it simple to perceive over inventory conditions accumulated every day, such as the distinction between a-1 and a-2. This example demonstrates an improvement result for "shipping following production" (similar to JIT). The meaning is production immediately took after by shipment. The situation before improvement was a vast gap amongst production and shipment, such as an abnormal state of inventory. A production plan was settled on by forecasting in light of past shipment records; this is the result of estrangement of production and shipment. This means unnecessary production was planned and made. Unnecessary work was finished. Another point is production lead time, such as b-1 and b-2.

Changing production A to B causes improvement by reducing inventory as well as by shortening lead time. Figure of accumulated chart practice demonstrates the practice of improvement results through accumulated.

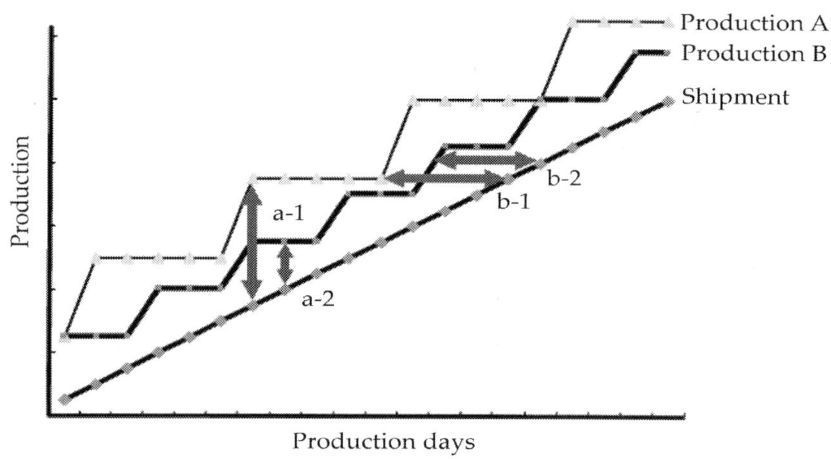

Figure 106 Accumulated chart (The Joint Development Board sponsored by the Association of Cost Engineers and the Royal Institute of Chartered Surveyors, n.d.)

Production and is before and production B is after improvement of shipping following production. The improvement results changed inventory level from a-1 to a-2, and production lead time from b-1 to b-2. Production after improvement is precisely following actual shipment.

Inventory reduction is anything but difficult to see as represented by the vertical bar; it demonstrates a reduction from 302 to 169 toward the finish of the month.

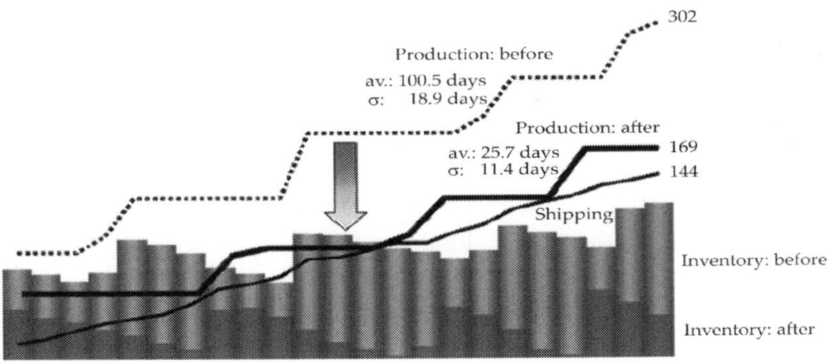

Figure 107 An accumulated chart use (The Joint Development Board sponsored by the Association of Cost Engineers and the Royal Institute of Chartered Surveyors, n.d.)

Then, production lead time diminished from a normal of 100.5 days to 25.7 days. Another issue of U with respect to productivity is that capacity depends on U. A similar example is the cost reduction per products base and company level profitability. An important point for the relation is the way to connect every product or parts cost reduction to the company's level of payment reduction. It's straightforward but quite often; companies make this sort of mistake.

A company may use a more elevated amount of manufacturing M and the P level may is likewise be adequate (100%), but the productivity results may in any case not contribute to the company's level of productivity without a more elevated amount of U of manpower or machine U for the correct level of inventory. Sometimes management is more intrigued by the percentage of machine utilization because of the high monetary investment. You shouldn't have any difficulty understanding "economical management" with the accompanying example. When you buy a luxury auto like a Rolls Royce, just keeping the auto in a carport will mean never driving it out and about. It is a cheap method because your contributed money sank the cost and devalued your investment. Figures mentioned below further illustrate the power of these management techniques. (Sakamoto, n.d.)

General improvement issues of U are:

- Tools and materials should be situated to settled regions and in suitable condition, including spare subjects.
- Fixed periodical maintenance of tools but machines should be set and kept in the best condition.
- Unavoidable turbulence should be accepted and prepared for with maintenance and machine utilization.
- Supplied components should be inspected preceding assembly.
- Practical scheduling depends on current production conditions.

Methods and Performance Meaning with Standard Time

Methods in engineering standard time include shop and/or workplace layout, number of laborers (manning), machine or office operating speed, sequences of operations, material handling, dances and fixtures, operation procedures, and motions (See Figure below). P includes lost time due to specialists, management, and necessary situations. Misfortunes in production stages are unreasonable interruption of work, low motivation, low effort level, deformity products due to inconsiderateness of laborers, slight for standard operation procedures, low ability level, and so on, these are specialists' responsibilities. It means the FM should supervise and instruct to improve/reduce specialists' misfortune time. These sorts of misfortunes are more than 80% of total misfortune time in shops.

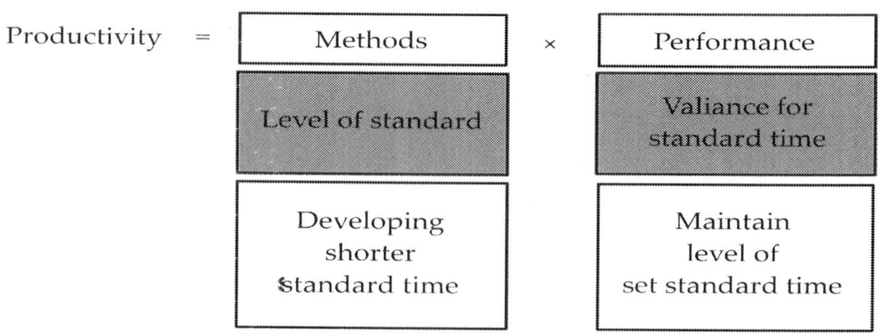

Figure 108: M and P meaning with time standard (Sakamoto, n.d.)

Next are management's or support staff's responsibilities such as sit time due to materials, machines/offices, and quality imperfections for out-of-control reasons. This is under 20%. The last one is unavoidable nonworking time such as power failure, mischance, catastrophe, and worker's organization action. These make up an insignificant size of nonworking hours. FMs' responsibility as to misfortune time should be controlled with a performance control system in light of designed time standard. Take a gander at Figure below in which laborers employ a couple of various methods before setting standard time. A standard method as a premise of standard time is the minimum time-consuming operation among practices at the shop. Then the methods change to time valued as standard time. This is not an improvement; it is welcomed as a change method. Thus, measure of method viability is the measured distinction of time value, and the most limited time value of method is normally the one most ideal way. Then, P is measured by laborer results to keep standard methods. Fluctuation of actual time value means contrast from standard methods in time value.

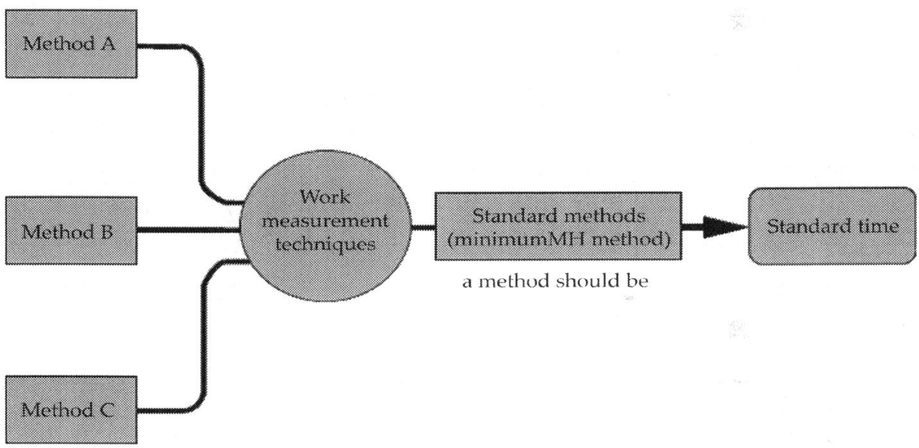

Figure 109: methods against standard time (Sakamoto, n.d.)

This is important to improve productivity in view of the engineering state of mind. Methods can evaluate those points of interest or detriments with a time value of standard time, whether those methods should be possible by the operator or not at shops by any means. Evaluation of methods should be possible in a cool or static approach to rate the conditions or results at shops. Methods change for productivity improvement is a movement to develop shorter standard time work methods. Along these lines, even the better methods with standard time evaluation don't take after standard time at shops sometimes. This is the reason laborers' P must be measured with standard time. P is measuring the contrast between actual time and standard time in percentages. A 100% P rating means the laborer took after standard time within the actual operation. Additionally, the laborer took after standard methods as a premise of standard time including standard pace of working or others that encompass the standard time "definition". A rating of lower than 100% for P means poor actual condition compared to standard time contents; more than 100% means actual operator's method was superior to anything standard time contents coupled with higher working pace. Productivity results can be measured with product of level/contrast of more than one method and actual after condition to standard time as P. Performance results can fluctuate depending on which method the laborer employs and whether he holds fast to standard time. (Sakamoto, n.d.)

B. Niebel stated: "Standards are the finish of time study or work measurement. This technique sets up a time standard permitted to perform a given undertaking, in light of measurements of the work content of prescribed methods" (Niebel and Freivalds 2003). There is a common approach to set targets and evaluate improvement, simply measure and compare two methods of the current improvement. There is no goal decency, as this is especially a subjective way. There is a long history of developing improved work measurement techniques while maintaining theoretical background for keeping reasonable accuracy and reasonableness (Bayha and Karger 1977).

Meaning of Standard Time on the Productivity Dimension

We should introduce an example of U in which there is a need for change over time that reduces production capacity and makes nonproduction time. Total changeover time per month, for example, is changeover or set-up time as a standard that takes after methods and number of occurrence per month. The method of changeover can reduce standard time with improvement of SMED, for example. This issue of changeover as the U dimension is evaluated by its impact as M dimension. Another part of changeover, total frequency of changeover, is the outcome due to production planning and control. Along these lines, it is possible to measure for the impact of total number of changeover times. There is no standard of a reasonable number of changeover times per month;

however, the desirable number of time is as small as possible. This is why a practical measurement of this matter of U dimension is measured with the difference between actual results of total changeover time and total changeover time in a benchmark month. It is possible to measure quality and facility maintenance matters with a standard time change. One operator added for standard manning is allocated in order to prevent quality problems, for example. Then allocated manning difference is possible to measure as standard, so the difference of that standard time is effect of this matter of productivity. Similar measures can be effective for preventive maintenance of machines and facilities. (Sakamoto, n.d.)

Let's introduce a new and meaningful relationship of M and P for productivity improvement. It is also a significant point of view for effective management. I can explain this relationship with common tools in kitchen use. Water is in a bamboo basket or metal bowl while in the water. Water leaks when the bamboo basket lifts up from water but there is no leaking from the metal bowl. There is a simple example of M & P. The basket or bowl is the method. Water gets into them when they are in the water but water leaks out with the bamboo basket. In order to keep water in the bamboo basket, it has to be sealed to protect against leaking water; this protective action is P. Water cannot be kept in the bamboo basket (M) without protection (P). Like this, no method can guarantee its (M) effect without protection (P) issues. Methods effectiveness occurs with measurement and evaluation with standard time whether meeting those time values or not. This is the difference of methods effectiveness and it is measured by standard time. The difference of standard time value is the difference of methods effectiveness. How much to meet those methods are measured as P level. Difference of P level is not dependent on methods themselves. This is why actual methods contribution such as MxP cannot manage without measuring P. (Sakamoto, n.d.)

U means taking water or not with a bowl or a protected bamboo basket. Doing an operation such as a method or not belongs in the U category for productivity. Before M and P, the necessity to do or not to do such a work has to be decided upon.

Figure below shows which working areas of a factory are effective to implement WIP. Each working area has its own cost of production, such as JPY 23,295 for a product at raw material stage and JPY 32,400 after press work. There is a common way of thinking today that minimizes the importance of production lot size. This is not a problem exactly, but there is a missing equal-value point of view. Six is set as the final assembly line, for example, and any working area is also set at six in lot size. There is no special reason to keep the same number of lot size for all working areas. At a metal workshop, 2–3 min are spent for punch press work for six pieces then set up/changeover every 15 or 20 min. Equal value for raw material is 12M JPY compared to a product 100M JPY, 22M JPY for after-press work. This means about ten times inventory for raw material is the equal value of five times before final assembly. (Sakamoto, n.d.)

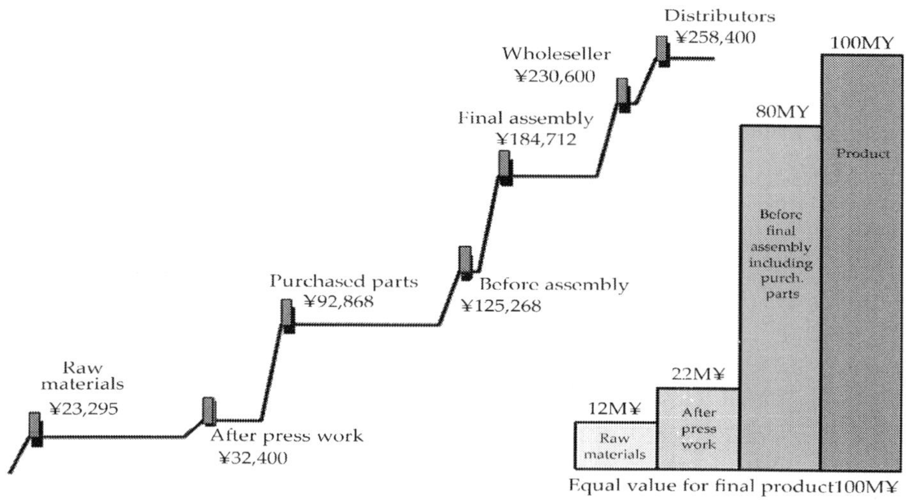

Figure 10.10: cost from raw materials for finishing (Sakamoto, n.d.)

Ten times for raw material, five times for before final assembly are acceptable sizes of inventory. The result is reducing lead time because there is enough WIP before final assembly, before press work and so on. Discovery of a solution of this kind belongs to the U dimension.

References

1. Adedeji Badiru, A.B.A.B., n.d. Industrial Project Management: Concepts, Tools, and Techniques. CRC Press.
2. Alistair Barros, A.G.E.K., ed., n.d. Business Process Management: 10th International Conference. Springer Shop.
3. Bessant, J.R., n.d. Managing advanced manufacturing technology: the challenge of the fifth wave. Blackwell Publishing Limited.
4. George A. Steiner, W.G.R., n.d. Industrial Project Managememt. Simon and Schuster.
5. Horngren, C.T., n.d. Cost Accounting: A Managerial Emphasis. Pearson Education India.
6. Sakamoto, S., n.d. Beyond World-Class Productivity: Industrial Engineering Practice and Theory. Springer Science & Business Media.
7. Antis WH, Honeycutt JR, Koch EN (1973) The basic motion of MTM, 4th edn. The Maynard
8. Foundation & Prentice-Hall, Upper Saddle River, NJ

9. Burnham DC (1972) Productivity improvement. Columbia University Press, New York
10. Carrol P (1954) Time study for cost control. McGraw-Hill, New York
11. Fried HO, Knox Lovell CA, Schmidt S (2008) The measurement of productive efficiency and productivity growth. Oxford University Press, New York
12. Fujita A (1953) Basics of industrial engineering. Kenpakuya, Tokyo, Japan
13. Gadiesh O, Gilbert JL (1998) Profit polls: A fresh look at strategy. Harvard Business Review,
14. May–June 1998
15. Herbert S (1971) The meaning and measurement of productivity. Bureau of Labor Statistics
16. Bulletin 1714
17. Honeycutt A, William JM, Kock EN (1973) The basic motion of MTM, 4th edn. The Maynard
18. Foundation & Prentice-Hall, Upper Saddle River, NJ
19. IMD, International Institute for Management Development (1997), (1998), (1999), The World
20. Competitiveness Yearbook, Lausanne, Switzerland
21. Institute of Industrial Engineers (1983), Industrial Engineering Terminology, Institute of Industrial Engineers, Norcross, GA
22. Juran JM (1995) Managerial breakthrough: The classic book on improving management performance. McGraw-Hill, New York
23. Kadota T, Sakamoto S (1992) Chapter 55: Methods analysis and design. In: Salvendy G (ed)
24. Handbook of industrial engineering. Wiley, New York, pp. 1415–1445
25. Krick EV (1965) An introduction to engineering & engineering design. John Wiley & Sons, New York
26. Lokiec M (1977) Productivity and incentives. Bobbin Publications, Los Angeles, CA
27. Mali P (1978) Improving total productivity. Wiley, NY

28. Meadow DH, Meadow DL, Randers J, Behrens WW III (1972) The limit to growth. Universe Books, New York
29. Morony MJ (1964) Facts from figures. Penguin Books, New York
30. Morrow RL (1957) Motion economy and work measurement. The Ronald Press, New York
31. Mundel M, Danner D (1994) Motion and time study improving productivity, 7th edn. PrenticeHall, Upper Saddle River, NJ
32. Nalebuff B, Brandenburger AM (1996) Co-opetition. Harper Collins Business, London, UK
33. Polanyi M http://infed.org/thinkers/polanyihtm
34. Prentice-Hall, Upper Saddle River, NJ
35. Prokopenko J (1987) Productivity management, a practical handbook. International Labour
36. Office, Geneva, Switzerland Riggs JL, Felix GH (1983) Productivity by objectives: Results-oriented solutions to the productivity puzzle. Prentice-Hall, Englewood Cliffs, NJ
37. Sakamoto S (1977) How a Japanese firm doubled productivity without capital investment. International Productivity Conference, Sydney, Australia
38. Sakamoto S (1977b) Japanese firm doubles productivity. Institute of Practitioners In: Work study, organization and methods. Management Services, UK
39. Sakamoto S (1981) Practices of industrial engineering. Kenpakusya, Tokyo, Japan
40. Sakamoto S (1983a) MOP: A head of OA, adopt IE to office. Annual Industrial Engineering Conference, Louisville, KY
41. Sakamoto S (1983b) Practices of work measurement. Japan Management Association, Tokyo, Japan
42. Sakamoto S (1985a) MDC engineering manual. Japan Management Association, Tokyo, Japan
43. Sakamoto S (1985b) MOP: Managing Office Productivity. Japan Management Association, Tokyo, Japan
44. Sakamoto S (1989) Process design concept. Ind Eng 3:31–34
45. Sakamoto S (1990) Really high Japanese productivity. Japan

Management Association, Tokyo, Japan

46. Sakamoto S (1991a) The MDC training manual. Productivity Partner Inc, Nara, Japan
47. Sakamoto S (1991b) MDC: Productivity engineering methods. Japan Management Association,
48. Tokyo, Japan
49. Sakamoto S (1992a) Design concept for methods engineering. In: Hodson WK (ed) Maynard industrial engineering handbook. McGraw Hill, New York
50. Sakamoto S (1992b) A practical manual of MDC. Japan Management Association, Tokyo, Japan
51. Sakamoto S (1997) Japanese firm doubles productivity, Management Services, Institute of Practioners in Work Study, Organization and Methods
52. Sakamoto S (2002) A study of company dignity. Toyokeizai Shinposya, Tokyo, Japan
53. Sakamoto S (2006) Methods design concept: An effective approach to profitability. J Philippine
54. Ind Eng
55. Sakamoto S (2007) Productivity management: Innovative approach for white color. Sangyou
56. Nouritsu University, Tokyo, Japan
57. Sakamoto S (2009) Return to work measurement. J Indust Engng 3:24
58. Schonberger RJ (1986) World-class manufacturing. The Free Press, New York
59. Skinner W (1978) Manufacturing in the corporate strategy. Wiley-Interscience, Hoboken, NJ
60. Slywotzke AJ, Morrison DJ (1997) The profit zone. Times Business, New York
61. Stockholm Environment Institute (1996) Sustainable economic welfare in Sweden: A pilot index
62. 1950–1992. Stockholm Environment Institute, Stockholm, Sweden
63. Swedish Federation of Productivity Services (1993) SAM training

program. Swedish Federation of Productivity Services, Stockholm, Sweden

64. Taylor FW (1911) The principles of scientific management. Harper, New York
65. Tiefenthal R (1975) Production: An international appraisal of contemporary manufacturing systems and the changing role of workers. McGraw-Hill, New York von Weizsäcker EU, Lovins AB, Lovins LH (1995) Faktor Vier. Rocky Mountain Institute, Boulder, CO
66. Zandin KB (1980) MOST work measurement system. Marcel Dekker, New York
67. Salvendy, G., n.d. Handbook of Industrial Engineering: Technology and Operations Management. Wiley.
68. The Joint Development Board sponsored by the Association of Cost Engineers and the Royal Institute of Chartered Surveyors, n.d. Industrial Engineering Projects: Practice and procedures. CRC Press.
69. Tonchia, S., 2008. Industrial Project Management: Planning, Design, and Construction. Springer Science & Business Media.
70. Waldemar Karwowski, G.S., n.d. Organization and Management of Advanced Manufacturing. Wiley.

Index

A

Activity control packages 89
Allied Quality Assurance Publications (AQAPs) 131
already expressed 102
announcement 18

B

beneficiary 62
Budget 47, 50

C

Call-off orders 115
Change control 24
commissioning 22, 27, 28
Conceptual design 22, 23, 43
construction 1, 2, 3, 4, 5, 7, 8, 9, 10, 11, 13
Cost Decrease Activity for the New Time 11
co-venturers 25
current international 146

D

drawings 91, 94, 99, 102, 103, 104, 107

E

elegant properties 137
engineering businesses 8
engineering ventures 53
Equipment Requires Inspection 190
Estimating data 188

F

four alpha numericals 93

G

General Electric 137, 152, 154

guarantee 3, 6, 11

H

Historically 208

I

illuminating prerequisites 114
insurance agency 116
internal handling help (IPR) 127
International Organization for Standardization (ISO) 154
IP belonging 206

J

just in time (JIT) 204

K

key deliverable 100
key management issues 203

L

logic network systems 68

M

management contractual 193
Material certificates 118
material got report (MRR) 128
merchandise got note (GRN) 128

N

networks 74, 75
network systems 68
noteworthy project 55, 57, 58
Nowadays 203

P

PC aided design (CAD) 103
planning action 65, 67
planning engineer 66, 67, 71, 75, 80
planning frameworks 67
Profitability development 207
project manager 13
project program 8

Q

quality affirmation and quality control (QA/QC) 8
quality function deployment (QFD) 143
Quality Management System (QMS) 149
quality plan 141, 143, 146
Quality System Regulation (QSR) 149

R

Reliability 132, 134

S

significance 65, 67
Situations 20
Six Sigma project 152, 153
solitary arrangement 158
spending estimate 46, 47, 50, 60, 61
structural engineering businesses 53

V

value engineering (VE) 167
value examination (VA) 167
value investigation 168
value management 168, 175

Value management (VM) 167

W

work break- down structure (WBS) 88
work breakdown structure (WBS) 89